Reliability Theory and Models

Stochastic Failure Models, Optimal Maintenance Policies, Life Testing, and Structures

Notes and Reports in Computer Science and Applied Mathematics

Editor
Werner Rheinboldt
University of Pittsburgh

1. JOHN R. BOURNE. Laboratory Minicomputing
2. CARL TROPPER. Local Computer Network Technologies
3. KENDALL PRESTON, JR. AND LEONARD UHR (Editors). Multicomputers and Image Processing: Algorithms and Programs
4. STEPHEN S. LAVENBERG (Editor). Computer Performance Modeling Handbook
5. R. MICHAEL HORD. Digital Image Processing of Remotely Sensed Data
6. SAKTI P. GHOSH, Y. KAMBAYASHI, AND W. LIPSKI (Editors). Data Base File Organization: Theory and Applications of the Consecutive Retrieval Property
7. ULRICH W. KULISCH AND WILLARD L. MIRANKER (Editors). A New Approach to Scientific Computation
8. JACOB BECK, BARBARA HOPE, AND AZRIEL ROSENFELD (Editors). Human and Machine Vision
9. EDGAR W. KAUCHER AND WILLARD L. MIRANKER. Self-Validating Numerics for Function Space Problems: Computation with Guarantees for Differential and Integral Equations
10. M. ABDEL-HAMEED, ERHAN ÇINLAR, AND JOSEPH QUINN (Editors). Reliability Theory and Models: Stochastic Failure Models, Optimal Maintenance Policies, Life Testing, and Structures

Reliability Theory and Models

Stochastic Failure Models, Optimal Maintenance Policies, Life Testing, and Structures

Edited by

Mohamed S. Abdel-Hameed

Department of Mathematics and Computer Science
University of North Carolina at Charlotte
Charlotte, North Carolina

Erhan Çınlar

Department of Industrial Engineering and Management Sciences
Northwestern University
Evanston, Illinois

Joseph Quinn

Department of Mathematics and Computer Science
University of North Carolina at Charlotte
Charlotte, North Carolina

1984

ACADEMIC PRESS, INC.

(Harcourt Brace Jovanovich, Publishers)

Orlando San Diego New York London
Toronto Montreal Sydney Tokyo

Academic Press Rapid Manuscript Reproduction

Proceedings of a Symposium on Stochastic Failure Models, Replacement and Maintenance Policies, and Accelerated Life Testing, Sponsored by the University of North Carolina at Charlotte and the U.S. Army Research Center, Held in Charlotte, North Carolina, June 24–26, 1983

ACADEMIC PRESS, INC.
Orlando, Florida 32887

United Kingdom Edition published by
ACADEMIC PRESS, INC. (LONDON) LTD.
24/28 Oval Road, London NW1 7DX

Library of Congress Cataloging in Publication Data

Main entry under title:

Reliability theory and models.

Includes bibliographical references.
1. Reliability (Engineering)--Congresses.
I. Abdel-Hameed, M. (Mohamed) II. Cinlar, E. (Erhan), date . III. Quinn, J.
TA169.R46 1984 620'.00452 84-45220
ISBN 0-12-041420-1 (alk. paper)

PRINTED IN THE UNITED STATES OF AMERICA

84 85 86 87 9 8 7 6 5 4 3 2 1

Ismail N. Shimi

1935–1982

Ismail N. Shimi, program manager for probability and statistics at the Air Force Office of Scientific Research, died suddenly on the morning of August 31, 1982. He was 47 years old.

He obtained his doctorate from the University of North Carolina in 1964. He taught at the University of California at Riverside, Ain Shams University in Cairo, and Florida State University at Tallahassee before joining the Air Force Office of Scientific Research in 1975.

His research work, which appeared in over 20 publications, covers most of applied probability: branching processes, traffic theory, reliability and life testing, stopping rules, maintenance, inventories, and so on. In addition, he was familiar with much of statistics and kept up with the latest developments in the theory of stochastic processes. This research experience, his taste for good mathematics, and his administrative skills were combined to make him an exceptional program manager. Under his leadership, the AFOSR program in probability and statistics became a most innovative, high-quality, modern research program. He had the rare ability to balance the short-term needs of his organization with long-term interests in basic research.

In 1976, together with C. P. Tsokos, he organized a conference on reliability theory and its applications. Its proceedings were published in two volumes by Academic Press in 1977. He was planning a similar venture when fate failed him. The present volume represents a partial fulfillment of his intentions. We dedicate it to his memory.

Contents

Life Testing

Reliability Structures, Computability, and Approximations

Special Session: U.S. Air Force Perspectives on Failure Models and Reliability/Availability/Maintainability

Contributors

Numbers in parentheses indicate the pages on which the authors' contributions begin.

Gary R. Adams (297), *New Engines Division, ASD/YZE, Wright-Patterson Air Force Base, Ohio 45433*

Harold S. Balaban (65), *ARINC Research Corporation, Annapolis, Maryland 21401*

R. E. Barlow (221), *Operations Research Center, University of California, Berkeley, Berkeley, California 94720*

H. W. Block (231), *Department of Mathematics and Statistics, University of Pittsburgh, Pittsburgh, Pennsylvania 15260*

Philip J. Boland (243), *Department of Mathematics, University College Dublin, Belfield, Dublin 4, Ireland*

Mark Brown (257, 267), *Department of Mathematics, City College, City University of New York, New York, New York 10021*

Erhan Çınlar (3), *Department of Industrial Engineering and Management Sciences, Northwestern University, Evanston, Illinois 60201*

Anthony J. Feduccia (295), *Systems Reliability and Engineering Branch, Rome Air Development Center, Griffiss Air Force Base, New York 13441*

Robert A. Fontenot (83), *Whitman College, Walla Walla, Washington 99362*

Guangping Ge (267), *Mathematics Department, Hebei Teachers' University, Shijiazhuang, Hebei, The People's Republic of China*

Gary Gottlieb (103), *New York University, New York, New York 10003*

Ina Parks S. Howell (199), *Department of Mathematical Sciences, Florida International University, Miami, Florida 33199*

Mei-Ling Ting Lee (273), *Bridgewater State College, Bridgewater, Massachusetts 02324, and Department of Mathematics, Boston University, Boston, Massachusetts 02215*

Benny Levikson (103), *Department of Statistics, Haifa University, Haifa, Israel*

D. T. McNichols (155), *Department of Statistics, Virginia Polytechnic Institute and State University, Blacksburg, Virginia 24061*

Richard J. Meinhold (169), *Department of Operations Research, George Washington University, Washington, D.C. 20052*

Francis J. O'Meara (292), *Operations Analysis Division, Science and Research (NR), HQ SAC, Offutt Air Force Base, Nebraska 68113*

W. J. Padgett (155, 177), *Department of Mathematics and Statistics, University of South Carolina, Columbia, South Carolina 29210*

Frank Proschan (83, 243), *Department of Statistics, Florida State University, Tallahassee, Florida 32306*

Joseph Quinn (115), *Department of Mathematics and Computer Science, University of North Carolina at Charlotte, Charlotte, North Carolina 28223*

T. H. Savits (231), *Department of Mathematics and Statistics, University of Pittsburgh, Pittsburgh, Pennsylvania 15260*

Margaret K. Schaefer (141), *Department of Mathematics, College of William and Mary, Williamsburg, Virginia 23185*

Moshe Shaked (43), *Department of Mathematics, University of Arizona, Tucson, Arizona 85721*

Nozer D. Singpurwalla (65, 169), *Departments of Operations Research and Statistics, George Washington University, Washington, D.C. 20052*

Robert T. Smythe (291), *Department of Statistics/Computer and Information Systems, George Washington University, Washington, D.C. 20052*

Preface

This volume collects most of the papers presented at a conference on reliability held at Charlotte, North Carolina, during June 24–26, 1983. All the papers have been refereed.

The aim of the conference was to bring together reliability theorists, statisticians, and experts on stochastic processes to discuss new and current directions for research on stochastic failure models, maintenance and replacement policies, and statistical and computational aspects of reliability.

One of the highlights of the meeting was a session titled "U.S. Air Force Perspectives on Failure Models and Reliability/Availability/Maintainability." The session consisted of presentations by representatives of three applied research groups in the Air Force. The presentations are included in this volume and are especially noteworthy for the glimpse they provide into the concerns of those dealing with the reliability and maintenance of very large complex systems.

The conference was supported by the University of North Carolina at Charlotte and the U.S. Army Research Center. We thank them for their support. We also thank M. Kazemi and Z. W. Ras for their time and effort with local arrangements. We owe much gratitude to Marty Grogg for her invaluable assistance both during the conference and throughout the preparation of these proceedings. Finally, we are especially grateful to the participants for a marvelous time and much fine mathematics.

Stochastic Failure Models

MARKOV AND SEMIMARKOV MODELS OF DETERIORATION

Erhan Cinlar[1]

IE/MS Department
Northwestern University
Evanston, Illinois

Lifetime distribution of a device is obtained assuming quite general laws for deterioration. Markov and semimarkov models of deterioration are discussed in the light of recent results on representation and characterization of Markov processes. Fairly general models can be obtained from a standard Poisson random measure by appropriate choices of several deterministic functions. The essential notion is a correspondence between the actual time (in actual use under field conditions) and an intrinsic time (in controlled laboratory conditions).

[1]Research supported by Air Force Office of Scientific Research through their grant AFOSR-82-0189.

ISBN 0-12-041420-1

1. INTRODUCTION

The simplest notion in reliability theory is the lifetime of a device, and the main problem connected with it is that of finding the distribution of the lifetime. In the statistical approach, that distribution is obtained from the experimental data on the lifetimes of a number of identical copies of the device. With high-reliability devices, the time and expense needed for such tests are prohibitive, and techniques like accelerated life testing and using censored data introduce further difficulties.

In an effort to circumvent such difficulties, the concept of a deterioration process is introduced, and the lifetime is defined to be the hitting time of a threshold value by the deterioration process. Then, one must deal with three separate problems: choosing an appropriate model for the deterioration process, solving for the hitting time distribution mathematically, and identifying the parameters of the model by statistical means. This paper is concerned with the first of these problems.

Our aim is to discuss the stochastic structure of some general models for deterioration processes and solve for the lifetime distributions in general. We move from the specific toward the general: we discuss deterioration processes that can be modeled by

a) continuous Markov

b) continuous semimarkov

c) right continuous Markov

d) Markov additive, and

e) semimarkov

processes. The last two are very close in structure, and together, they cover all processes proposed in the past literature as models for deterioration processes.

Recent representation and characterization theorems obtained in ÇINLAR [5] and in ÇINLAR and JACOD [6] for Markov processes enable us to represent the processes involved in terms of several deterministic functions. These functions have certain physical meanings and can be determined by laboratory experiments under controlled situations (as opposed to actual use under field conditions). The essential notion is a correspondence between the actual time (in actual use under field conditions) and an intrinsic clock time constructed so that the deterioration process appears smoother and simpler in clock time.

Section 2 is on continuous Markov and semimarkov models of deterioration; this is a simplified version of the results in [5] and provides insight into the techniques to follow. Section 3 describes the structure of increasing Markov processes, obtained in ÇINLAR and JACOD [6], in the setting of deterioration processes. Section 4 constructs a Markov additive process, which is a two-dimensional process (Y,A), Y being the deterioration process in intrinsic clock time and A being the actual time as a function of clock time. Finally, Section 5 constructs the semimarkov process modeling the deterioration process in actual time, and shows that the process is almost the most general possible. The paper ends with Section 6 computing the distribution of the lifetime under a theshold mechanism for failure.

This is an expository paper, written in a non-technical style with minimal notation, concentrating on modeling issues and on meanings to be attached to various parameters and operations. To keep it simple, we limited the exposition to the deterioration of a single component, but multi-component versions are almost as easy and can be constructed by mimicking the procedure here and consulting the sources referenced.

Notation and Preliminaries

Throughout the paper, x positive means $x \geqq 0$ and f increasing means $f(x) \leqq f(y)$ if $x \leqq y$; we add the term "strict" if strict inequalities hold; for instance, f is strictly increasing if $f(x) < f(y)$ whenever $x < y$.

Throughout, (Ω, A, P) will be a complete probability space; and the deterioration process will be denoted by $X = (X_t)$, X_t being the deterioration level at time t. We assume that $t \to X_t$ is increasing, real-valued, with $\lim_{t\to\infty} X_t = \infty$. In addition to the random variables X_t, there will be other random variables and processes defined on (Ω, A, P).

We assume that there is an increasing right-continuous augmented family (H_t) of sub-σ-algebras of A, and interpret H_t as the "history" until t of the process X and other auxiliary processes to appear. Finally, it will be convenient to write P_x for the conditional probability given that X_0 is x, and then, we write E_x for the corresponding conditional expectation operator.

2. MARKOV AND SEMIMARKOV MODELS OF WEAR

By a wear process we mean a positive increasing continuous stochastic process $X = (X_t)$. Recall the convention that $X_\infty = \lim_{t\to\infty} X_t = \infty$. The usefulness of such processes was pointed out in [8]. The present section is a simplified version of the detailed results of [5].

Suppose that our intuition regarding the wear process leads us to assume that X is a strong Markov process, that is, for any stopping time T of the history (H_t), the past H_T until T and the future $(X_{T+t})_{t\geqq 0}$ are conditionally independent given the present state X_T, or more precisely,

$$E_x[\, W \circ \theta_T \mid H_T] = E_{X_T}[W] \tag{2.1}$$

for every $x \geq 0$ and every positive measurable functional W of the process X, where $W \circ \theta_T$ is the same functional evaluated over the shifted path $t \to X_{T+t}$ (for instance, if $W = f(X_t, X_u)$ then $W \circ \theta_T = f(X_{T+t}, X_{T+u})$).

The following shows that our intuition must be faulty: all such processes are devoid of probabilistic interest.

(2.2) THEOREM. Let X be a positive increasing continuous strong Markov process. Then, there is a deterministic continuous strictly increasing function c from $\mathbb{R}_+$ to $\mathbb{R}_+$ such that

$$X_t(\omega) = c(c^{-1}(x)+t), \quad t \geq 0,$$

for almost every outcome $\omega \in \Omega$ with $X_0(\omega) = x$; (see Figure 1).

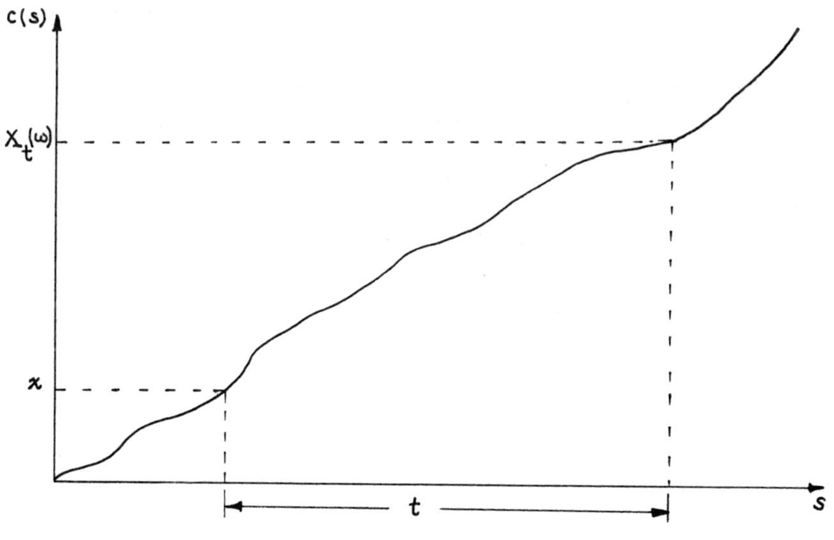

FIGURE 1. If the initial state is x, the Markov process must follow the path described by c, starting at position x.

PROOF. Define

$$C_t = X_t - X_0, \quad A_u = \inf\{t: C_t > u\}. \tag{2.3}$$

For each $u \geqq 0$, A_u is a stopping time, and by the continuity of (C_t),

$$X(A_u) - X_0 = C(A_u) = u. \tag{2.4}$$

a) Let $T = A_u$ for some fixed u, and write $W = I_{\{A_v \leqq t\}}$ for fixed v and t. Note that, in the notation of (2.1), $W \circ \theta_T = I_{\{A_{u+v} - A_u \leqq t\}}$ in view of (2.3) and (2.4). Hence, by the strong Markov property (2.1),

$$P_x[A_{u+v} - A_u \leqq t \mid \mathcal{H}_{A_u}] = P_{X(A_u)}[A_v \leqq t] = P_{x+u}[A_v \leqq t]. \tag{2.5}$$

In particular, this means that $A_{u+v} - A_u$ is independent of $(A_s)_{s \leqq u}$ given $X_0 = x$, that is, the process (A_u) has independent increments once X_0 is known.

b) Since X is increasing, continuous, and strong Markov, X must be strictly increasing. To see this, let $\varepsilon > 0$ be arbitrary and define

$$R_\varepsilon = \inf\{t: X_t = X_{t-\varepsilon}\}, \quad S_\varepsilon = \inf\{t > R_\varepsilon: X_t \neq X_{R_\varepsilon}\}.$$

Both R_ε and S_ε are stopping times, and by the continuity of X, we have $X(R_\varepsilon) = X(S_\varepsilon)$. Thus, by the strong Markov property, the future after R_ε has the same probability law as the future after S_ε. But, X is increasing immediately to the right of S_ε whereas X remains constant for some time after R_ε by the way R_ε and S_ε are defined. Hence, the strong Markov property can hold only if $R_\varepsilon = S_\varepsilon = \infty$. Since ε is arbitrary, this shows that X never remains constant, that is, X is strictly increasing.

c) Since X is strictly increasing, so is (C_t), and therefore the functional inverse (A_u) of (C_t) is continuous. Since X is continuous, (A_u) is strictly increasing. By part (a), A has independent increments. But, every continuous strictly increasing process with independent increments is deterministic. Hence, we must have

$$A_u(\omega) = A(x,u) \quad \text{if} \quad X_0(\omega) = x \tag{2.6}$$

for some deterministic function $A(x,u)$. Now (2.5) shows that

$$A(x,u+v) - A(x,u) = A(x+u,v),$$

which implies that (if we put $a(u) = A(0,u)$),

$$A(u,v) = a(u+v) - a(u), \quad u,v \geqq 0$$

for some strictly increasing continuous deterministic function a with $a(0) = 0$. Now, (2.6) becomes

$$A_u(\omega) = a(u+x) - a(x) \quad \text{if} \quad X_0(\omega) = x,$$

and (2.4) becomes

$$X(a(x+u) - a(x), \omega) = x + u \quad \text{if} \quad X_0(\omega) = x. \tag{2.7}$$

Define c to be the functional inverse of a; then,

$$a(x+u) - a(x) = t \Longleftrightarrow x + u = c(a(x)+t),$$

and (2.7) can be re-written as

$$X_t(\omega) = c(a(x)+t) = c(c^{-1}(x)+t), \quad X_0(\omega) = x,$$

as claimed. That c is strictly increasing and continuous follows from the fact that a is so. □

Note that, as shown in part (b) of the proof above, every increasing continuous strong Markov process is in fact strictly increasing. On this score alone, we can reject such processes as models for wear processes by noting that most devices experience periods of no wear corresponding to periods when the device is not being used. The next model is, therefore, more appropriate since it will be assumed that the strong Markov property holds only when the device is working.

Continuous Semimarkov Models of Wear

We are still assuming that the wear process $X = (X_t)$ is positive, increasing, and continuous, and that $X_\infty = \infty$. Let

$$\mathbb{M}(\omega) = \{t: X_{t+\varepsilon}(\omega) > X_t(\omega) \quad \text{for every} \quad \varepsilon > 0\}; \tag{2.8}$$

this is the set of times on which $t \to X_t(\omega)$ is increasing to the right. It can be obtained by removing from $\mathbb{R}_+$ all maximal intervals of the form $[a,b)$ on which $t \to X_t(\omega)$ remains constant.

(2.9) DEFINITION. The process X is said to be *semimarkov* provided that the strong Markov property (2.1) holds for every stopping time T of $(\mathcal{H}_t)$ such that $T(\omega) \in \mathbb{M}(\omega)$ for almost every ω with $T(\omega) < \infty$.

Thus, saying that the wear process is semimarkov amounts to saying that the future after T and the past until T are conditionally independent given X_T provided that the stopping time T is a "working" time. This allows the law governing the idle periods to be totally arbitrary.

WARNING. In much of the elementary literature, the term "semimarkov" is used for a very special and trite class of processes. The present usage is a generalization of the concept embedded in Lévy's original work [11]. See also JACOD [10] and Section 5 here.

(2.10) THEOREM. Let X be positive, increasing, continuous, and semimarkov. Then, $t \to X_t$ is the functional inverse of a strictly increasing right-continuous positive process with independent increments (generally non-stationary).

PROOF. Let (C_t) and (A_u) be defined as before, by (2.3). See Figure 2. Each A_u is a stopping time and $A_u(\omega) \in \mathbf{M}(\omega)$ for every ω. Thus, the strong Markov property holds at A_u, and the part (a) of the proof of (2.2) goes through. So, (A_u) has

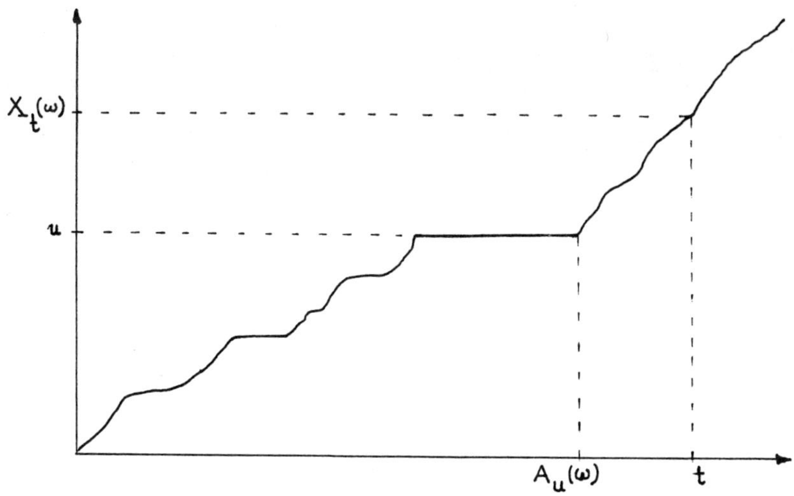

FIGURE 2. The processes (X_t) and (A_u) are functional inverses of each other when $X_0 = 0$; and (A_u) has independent increments.

independent increments. Since X is continuous, (A_u) is strictly increasing. Since X is increasing, (A_u) is right-continuous. Now, (C_t) is the functional inverse of (A_u) and $X_t = X_0 + C_t$. □

The structure of an increasing right-continuous process with independent increments is well-known (see ITO [9] for instance). We have

$$A_u = a(u) + \sum_{u_i \leq u} W_i + \int_{[0,u] \times \mathbb{R}_+} t\, N(dv,dt) \tag{2.11}$$

where

a) a is an increasing continuous deterministic function;

b) the u_i are some fixed points in $\mathbb{R}_+$, and the W_i are independent strictly positive random variables;

c) N is a Poisson random measure on $\mathbb{R}_+ \times \mathbb{R}_+$ independent of (W_i).

This yields the following explanation for the wear process (X_t) assuming that $X_0 = 0$ and $X = \lim_{t\to\infty} X_t = \infty$. The wear process X follows the deterministic path $t \to c(t) = a^{-1}(t)$ except for various interruptions during which X remains constant. The interruptions are of two kinds: scheduled and unscheduled. The scheduled interruptions occur at times when the wear level reaches pre-fixed levels $u_1, u_2, \ldots$ and the interruption at level u_i lasts a random time W_i. Unscheduled interruptions happen according to a Poisson like process and last arbitrary random amounts. These may be looked upon as failure times for the machine to which the device is attached (since such failures do not increase the deterioration of the device).

The probability law of the process (A_u), and therefore of (X_t), is described by the following objects:

i) a deterministic increasing continuous function a;

ii) some fixed points u_i,

iii) for each i, a distribution function ϕ_i;

iv) a σ-finite measure n on $\mathbb{R}_+ \times \mathbb{R}_+$ satisfying

$$n(\{u\}\times\mathbb{R}_+) = 0, \qquad \int_{[0,u]\times\mathbb{R}_+} n(dv,dt)\ \min(t,1) < \infty.$$

Here, ϕ_i is the distribution of W_i and may depend on u_i, and n is the mean measure of N, that is,

$$P[\mathbb{N}(B) = j] = \frac{e^{-n(B)}\, n(B)^j}{j!}, \quad j = 0,1,2,... \tag{2.12}$$

for every Borel subset B of $\mathbb{R}_+ \times \mathbb{R}_+$ with $n(B) < \infty$. Incidentally, (2.12) characterizes the Poisson random measure: if (2.12) holds for every Borel set $B \subset \mathbb{R}_+ \times \mathbb{R}_+$, then $N(B_1)$ ..., $N(B_k)$ are independent Poisson random variables whenever $B_1, \ldots, B_k$ are disjoint.

3. MARKOVIAN MODELS OF SHOCK AND WEAR

By a shock and wear process we mean a positive, increasing, and right-continuous stochastic process (X_t). We interpret X_t as the cumulative deterioration level of the device at time t and we think of a jump of (X_t) as the damage caused by a shock. As before, we assume $X_\infty = \infty$.

We have seen in Theorem (2.2) that, if X is continuous and strong Markov, then X is deterministic. Keeping the assumption of continuity and relaxing the strong Markov property yields the semimarkov model analyzed in Theorem (2.10) and the discussion following it. In this section we put back the strong Markov property and relax the continuity assumption. The following basic representation theorem was obtained in ÇINLAR and JACOD [6]. The proof is too long and complicated to be given here.

(3.1) THEOREM. Let X be a positive, increasing, right-continuous, and quasi-left-continuous strong Markov process. Then, there exist deterministic Borel functions $a: \mathbb{R}_+ \to (0,1]$, $b: \mathbb{R}_+ \to [0,1]$, $k: \mathbb{R}_+ \times \mathbb{R}_+ \to \mathbb{R}_+$, and a Poisson random measure N on $\mathbb{R}_+ \times \mathbb{R}_+$ with mean measure $n(ds,dz) = ds \cdot dz/z^2$ such that X is obtained from a process Y satisfying

$$Y_u = Y_0 + \int_0^u b(Y_s)\, ds + \int_0^u \int_0^\infty k(Y_{s-},z)\, N(ds,dz) \tag{3.2}$$

by a random time change

$$X_t = Y_{C_t}, \quad t \in \mathbb{R}_+, \tag{3.3}$$

where the clock-time process (C_t) is the functional inverse of the strictly increasing continuous additive functional (A_u) of Y defined by

$$A_u = \int_0^u a(Y_s)\, ds, \quad u \in \mathbb{R}_+. \tag{3.4}$$

This theorem is worth examining carefully. The remainder of this section is devoted to explaining it, its limitations, and its extensions.

Hunt and Ito Processes

It is useful to introduce here some terms for the sake of brevity. A process taking values in $\mathbb{R}$ (or $\mathbb{R}^n$) is said to be a Hunt process provided that it be right-continuous, quasi-left-continuous, and strong Markov. A Hunt process is said to be an Ito process if it satisfies an integral equation like (3.2).

Quasi-left-continuity refers to the property that $X_{T_n} \to X_T$ almost surely for any increasing sequence of stopping times T_n with limit T. It is a stronger version of stochastic continuity; it implies that the probability of a jump occuring at a pre-fixed time t is zero, and more generally, that the jumps of the process cannot be predicted before they occur. A Poisson process is a good example: it is full of jumps but no jump can be predicted.

With these terms in place, we may re-phrase Theorem (3.1) as follows: *every increasing Hunt process is obtained from an increasing Ito process by a random time change.*

Role of the Time Change

The processes (C_t) and (A_u) are functional inverses of each other. They set up a correspondence between the actual time and an intrinsic clock time: when the clock shows u, the actual time is A_u; when the actual time is t, the clock shows C_t. The time scale for the Ito process (Y_u) is the clock time, whereas for the original Hunt process (X_t), the time scale is the actual time. Note that both (A_u) and (C_t) are strictly increasing and continuous, and

$$X_t = Y_{C_t}, \quad Y_u = X_{A_u}. \tag{3.5}$$

Thus, through the time change, the actual deterioration process is smoothed out and the shocks are made into the points of a Poisson random measure. It is best to think of C as a clock attached to the device and whose mechanism is affected by the device in such a manner that, when the time is reckoned by using that intrinsic clock, the observed process (Y_u) is better behaved and easier to describe. The relationship between various processes involved are shown in Figure 3 below.

Shock and Wear Process in Clock Time

This is the process (Y_u). It is a Hunt process and an Ito process. The right side of (3.2) shows its wear and shock components separately. The first integral there is continuous and increasing in u and may be interpreted as the amount of wear experienced until the *clock* shows u. Thus, $b(x)$ is the rate of wear per unit of clock time when the deterioration level is x. The second integral is in fact a sum: if the points of the Poisson random measure are enumerated so that

$$N(B) = \sum_i 1_B(U_i, Z_i), \quad B \subset \mathbb{R}_+ \times \mathbb{R}_+ \quad \text{Borel,} \tag{3.6}$$

then

$$\int_0^u \int_0^\infty k(Y_{s-}, z)\, N(ds, dz) = \sum_{U_i \leq u} k(Y_{U_i -}, Z_i). \tag{3.7}$$

We may think of each point (U_i, Z_i) as representing a shock; U_i is the clock time of the ith shock, and Z_i is a magnitude associated with it. Then, (3.7) and (3.2) imply that, if a shock

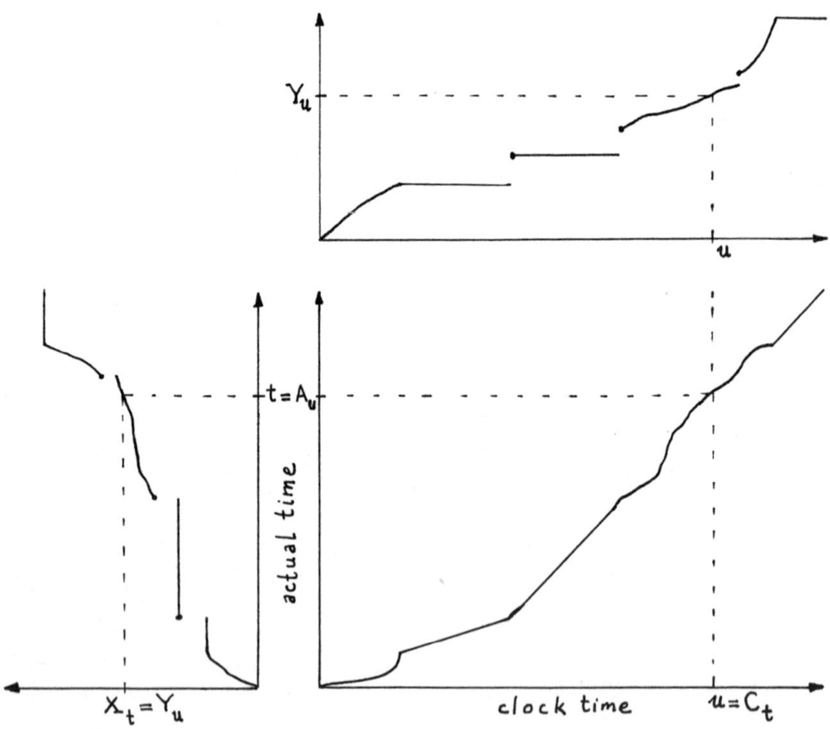

FIGURE 3. The deterioration process X runs on an actual time. It appears as the process Y when viewed in clock time. The clock runs so that the actual time passes at the rate a(x) per unit of clock's time while the deterioration level is x.

occurs at some time u with "magnitude" z, and if the deterioration level just before the shock was $Y_{u-} = x$, then a damage of size $k(x,z)$ is experienced at that time u.

The equation (3.2) may thus be summarized as follows. The device is operating in a shock producing medium. The "times" and "magnitudes" associated with different shocks form a collection of random points (U_i, Z_i) in $\mathbb{R}_+ \times \mathbb{R}_+$. The number $N(B)$ of such pairs (U_i, Z_i) falling in a Borel set B has the Poisson distribution with mean equal to $n(B) = \int_B ds\, dz/z^2$. Each shock causes some damage, possibly none: if the shock occurs when the deterioration level is x and if the shock "magnitude" is z, then the damage to the device is $k(x,z) \geq 0$. In addition to

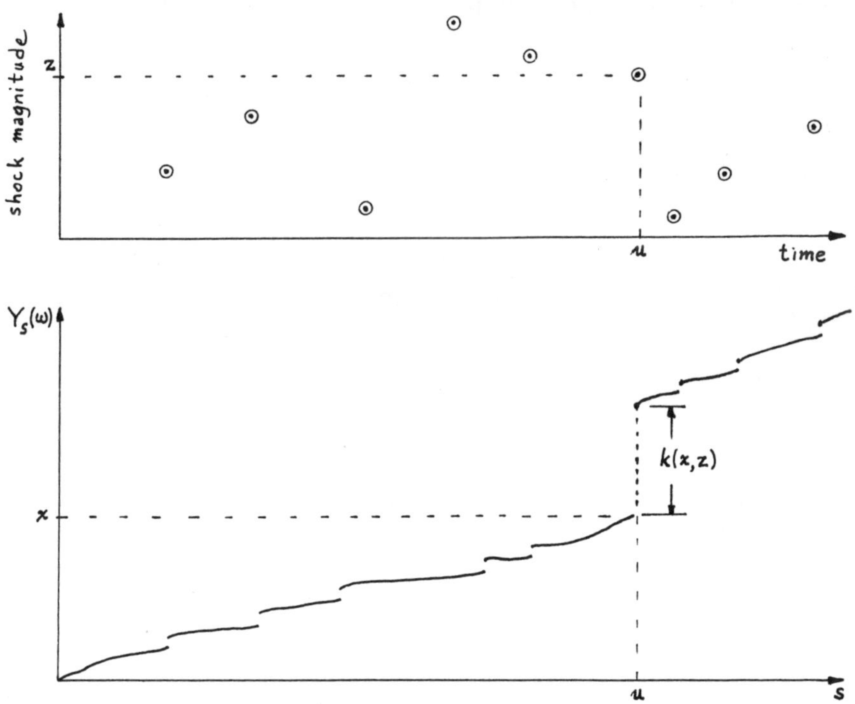

FIGURE 4. Damage caused by a shock of magnitude z is $k(x,z)$ if the deterioration level just before the shock is x.

damages caused by shocks, the device experiences "wear" at the rate $b(x)$ per unit "time" when the deterioration level is x. See Figure 4 for a pictorial re-statement.

Shock Mechanism

This concerns the Poisson random measure N. As we have argued above, N is a counting measure and its points may be thought of as representing shocks. However, the exact meanings to be associated with "time" and "magnitude" need to be considered carefully. One role played by the random time change was that the shocks occur homogeneously in clock time; we leave this as it is. But the term "magnitude" for a shock is rather ill-defined. Recall that, in Theorem (3.1), N was taken standardized so that the expected number of shocks by clock time u whose "magnitudes" belong to the Borel set $A \subset \mathbb{R}_+$ is

$$E[N([0,u]\ A)] = u \int_A \frac{1}{z^2}\, dz. \tag{3.8}$$

It is obvious that this is a highly stylized choice.

(3.9) *Theorem* (3.1) *remains true when* N *is replaced by any Poisson random measure* $\hat{N}$ *on* $\mathbb{R}_+ \times \mathbb{R}_+$ *with mean measure*

$$\hat{n}(du,dz) = du \cdot \nu(dz),$$

where ν *is an arbitrary* σ*-finite, infinite, diffuse measure on* $\mathbb{R}_+$*, provided that* k *is replaced by an appropriate function* $\hat{k}$.

For most practical purposes, the rate $\nu(dz)$ of shocks of "magnitude" z is very large for z small and is very small for z large. Therefore $\nu(dz) = dz/z^2$ is a realistic measure for this purpose. Thus, the correct interpretation of "magnitude" of a shock is bound up with the choice of the measure ν in statement (3.9), and the correct k is to be chosen after the meaning for shock magnitude is settled.

It is possible that $k(x,z) = 0$ for some x and z. Thus, some shocks might not cause any damage. A global theorem like (3.1) foresees all possibilities, but the representation for (Y_u) can be simplified in specific instances. For instance, if the expected number of shocks per unit time is bounded by, say 1, it is sufficient to replace N by the Poisson random measure $\hat{N}$ on $\mathbb{R}_+ \times [0,1]$ with mean measure $n(ds,dz) = ds \cdot dz$ (then, the U_i form a simple Poisson point process on $\mathbb{R}_+$ and the Z_i are independent and uniformly distributed). In the representation (3.2) this can be affected by simply setting $k(x,z) = 0$ for all $z \leqq 1$ and all $x \geqq 0$.

Warning

Even though the shock mechanism is Poisson, the jump times of $u \to Y_u$ do not form a Poisson process in general. If the jump times of $u \to Y_u$ are to form a Poisson process with rate λ, we can handle it by making sure that $k(x,z) > 0$ for all $z \geqq 1/\lambda$ and all $x \geqq 0$ and setting $k(x,z) = 0$ for all $z < 1/\lambda$ and all $x \geqq 0$.

Computations with Y

The form of (3.2) is especially convenient as a starting point for computations. For any differentiable function f, a change of variable argument (this is a generalization of the usual change of variable formula in calculus and is a specialization of Ito's formula) yields

$$f(Y_u) = f(Y_0) + \int_0^u f'(Y_s)\, b(Y_s)\, ds + \sum_{s \leq u} [f(Y_s) - f(Y_{s-})]$$

$$= f(Y_0) + \int_0^u f'(Y_s)\, b(Y_s)\, ds + \int_0^u \int_0^\infty [f(Y_{s-} + k(Y_{s-}, z)) - f(Y_{s-})]\, N(ds, dz),$$

where f' denotes the derivative of f. Here u is arbitrary and can be replaced by random times as well. Taking expectations on both sides, and using the fact that $N(ds,dz) - ds\, dz/z^2$ is a martingale measure, gives

$$E_x[f(Y_u)] = f(x) + \int_0^u E_x[f'(Y_s)\, b(Y_s)]\, ds + E_x\left[\int_0^u \int_0^\infty [f(Y_s + k(Y_s, z)) - f(Y_s)]\, ds\, \frac{dz}{z^2}\right]. \tag{3.11}$$

Introducing the operator

$$Gf(x) = f'(x)\, b(x) + \int_0^\infty [f(x + k(x,z)) - f(x)]\, \frac{dz}{z^2}, \tag{3.12}$$

the equation (3.11) above becomes

$$E_x[f(Y_u)] = f(x) + E_x \int_0^u Gf(Y_s)\, ds. \tag{3.13}$$

Finally, introducing the transition function

$$P_t(x, A) = P_x[Y_t \in A], \qquad x \geq 0, \qquad A \subset \mathbb{R}_+ \text{ Borel}, \tag{3.14}$$

and defining

$$P_t f(x) = \int_0^\infty P_t(x, dy)\, f(y), \tag{3.15}$$

the formula (3.13) becomes

$$P_u f(x) = f(x) + \int_0^u P_s Gf(x)\, ds. \tag{3.16}$$

This is the integrated form of Kolmogorov's forward equations. In particular, this shows that the operator G defined by (3.12) is the infinitesimal generator of the process Y, and its domain contains all differentiable bounded functions.

Uniqueness and Inverse Problems

It is clear by now that the fundamental process involved is the Ito process Y. For many practical situations, Y may provide a good enough model, which amounts to saying that the clock time and actual time are the same.

Theorem (3.1) assures us that, for any increasing Hunt process X, it is possible to define a clock time process which transforms X into a process Y satisfying (3.2) However, the equation (3.2) may have more than one solution unless b and k have certain smoothness properties. At present, there are no known conditions that are necessary and sufficient for uniqueness.

However, some conditions are known that ensure the existence and uniqueness of a solution Y to (3.2). Drosen [7] has shown that, if $b = 0$ and

$$k(x,z) \leqq k_0(z), \quad z \geqq 0,$$

for all x for some function k_0 satisfying

$$\int_0^\infty (k_0(z) \wedge 1) \frac{1}{z^2}\, dz < \infty,$$

then (3.2) has a solution Y. The same can be concluded with the extra condition that $x \to k(x,z)$ is increasing even if $b \neq 0$ but is such that the deterministic equation

$$y_t = y_0 + \int_0^t b(y_s)\, ds$$

has a solution. Uniqueness of solutions is even harder; generally one needs conditions like Lipschitz continuity on $x \to b(x)$ and $x \to k(x,z)$ in order to have a unique solution.

Extensions

Theorem (3.1) remains true for any Hunt process X taking values in $\mathbb{R}^n$ and whose paths are of finite variation, and then, the functions b and k are defined on $\mathbb{R}^n$ and $\mathbb{R}^n \times \mathbb{R}_+$ respectively and take values in $\mathbb{R}^n$. In the case of X taking values in $\mathbb{R}^n$, these provide models of self-repairing systems. In the case of X taking values in $\mathbb{R}^n$, a particularly interesting case is when each component of X is increasing; then, X is the most general Markovian model for a system of n components whose deterioration processes interact.

There are further extensions, which seem to be of little interest in reliability, to arbitrary semimartingale Hunt processes on abstract spaces.

4. MARKOV ADDITIVE PROCESSES

The main purpose of this section is to serve as an interlude between the preceding section and the next. However, the model here may be of interest in itself; see Interpretation (4.3) below, and the detailed account and references given in [4].

Our starting point is the Ito process Y examined in the preceding section. As before, we interpret Y_u as the deterioration level when the clock time is u. From this, the deterioration process X in actual time was obtained by a random time change, interpreting the additive functional (A_u) defined by

(3.4) as the actual time as a function of the clock time u. Our objective is to replace the actual time process by a more realistic one that allows the actual time to increase while the clock time remains stationary. This will allow for such things as repair times and unscheduled interruptions, during which the device does not work and does not deteriorate, and therefore, the clock time does not change.

Assuming that the path $u \to Y_u$ is known, the conditional structure of (A_u) will be that of a process with independent increments (like (A_u) in Section 2, but slightly less general):

$$A_u = \int_0^u a(Y_s)\, ds + \int_0^u \int_0^\infty g(Y_{s-},w)\, M(ds,dw) + \int_0^u \int_0^\infty \int_0^\infty h(Y_{s-},z,w)\, \hat{N}(ds,dz,dw), \qquad (4.1)$$

where

i) $a\colon \mathbb{R}_+ \to [0,1]$, $g\colon \mathbb{R}_+ \times \mathbb{R}_+ \to \mathbb{R}_+$, and $h\colon \mathbb{R}_+ \times \mathbb{R}_+ \times [0,1] \to \mathbb{R}_+$ are deterministic Borel functions;

ii) M is a Poisson random measure on $\mathbb{R}_+ \times \mathbb{R}_+$, independent of $\hat{N}$, with some mean measure $ds\, \mu(dw)$;

iii) $\hat{N}$ is a Poisson random measure on $\mathbb{R}_+ \times \mathbb{R}_+ \times [0,1]$, with mean measure $ds \cdot dz \cdot dw/z^2$; it is related to the Poisson random measure N figuring in (3.2) by

$$\hat{N}(ds,dz,[0,1]) = N(ds,dz), \qquad s \geqq 0,\ z \geqq 0.$$

The first integral in (4.1) defines a continuous additive functional of Y; it describes the continuous increase of actual time as a function of the clock time: the actual time increases at rate $a(x)$ per unit of clock time when the deterioration level is x.

The second integral in (4.1) depicts unscheduled interruptions. It is governed by a Poisson random measure M independent of the process Y. The points of M can be interpreted as "shocks" that do not affect the device but cause interruptions to the working of the device. If such a "shock" occurs at clock-time s and has intensity w, the work is stopped for a time $g(x,w)$ depending on the intensity w and the level x of deterioration.

The last term in (4.1) describes the effect of a system of shocks that cause damage to the device as well as interrupting the work. For this purpose, such a shock is represented by three indicators: the time s, the damage indicator z, and the interruption indicator w. As we have seen in Section 3, then, the damage to the device is $k(x,z)$ if $Y_{s-} = x$. Here, we further have an interruption lasting $h(x,z,w)$ units of actual time. This part is equivalent to saying that every time a shock occurs causing damage, an interruption of some random length occurs; the conditional distribution of interruption is some distribution $t \to H(x,y,t)$ if the damage takes the deterioration level from x to $x + y$. The function h is related to H through

$$h(x,z,w) = \inf\{t: H(x,k(x,z),t) > w\}.$$

(4.2) THEOREM. Let Y be an Ito process satisfying (3.2), and let A be as defined by (4.1) and such that

$$A_u < \infty, \quad u \geqq 0; \quad \lim_{u\to\infty} A_u = \infty. \tag{4.3}$$

Then, the two-dimensional process (Y_u, A_u) is a Markov additive process in the sense of [2], that is,

$$P_x[Y_{u+v}\in B, A_{u+v}-A_u\in C \mid Y_s, A_s;\ s\leqq u] = P_{Y_u}[Y_v\in B, A_v\in C] \tag{4.4}$$

for all $x,u,v \geqq 0$ and all Borel subsets B and C of $\mathbb{R}_+$.

(4.5) REMARK. In fact, (Y_u, A_u) is a strong Markov additive process: a statement similar to (4.4) holds with u replaced by a stopping time U of the process (Y_u, A_u).

(4.6) ALTERNATIVE INTERPRETATION. Markov additive processes of this generality were first introduced in [2]. In fact, the definition of [2] allows for even greater generality: the process Y can be an arbitrary standard Markov process, in the sense of BLUMENTHAL and GETOOR [1], taking values in an abstract space, and A can take values in $\mathbb{R}^n$ and have a more intricate structure. Such a general Markov additive process can be used as a totally different model for deterioration processes: we may interpret Y_u as the state of an "environmental process" at time u for an n-component device. This point of view was presented in some detail in [4]; see [2] and [3] for further details.

Potentials and Generators

The potential kernel of (Y_u, A_u) is defined by

$$U(x, B\times C) = E_x\left[\int_0^\infty 1_B(Y_u)\, 1_C(A_u)\, du\right] \tag{4.7}$$

for $x \geqq 0$, B and C Borel subsets of $\mathbb{R}_+$. Note that this is the expected amount of time spent by (Y_u, A_u) in the rectangle $B \times C$ given that $Y_0 = x$. If either B or C is bounded, then this expectation is finite for all x. For each x, (4.7) defines a measure $U(x,\cdot)$, and

$$\int_0^\infty\int_0^\infty U(x,dy,dt)\, F(y,t) = E_x\left[\int_0^\infty F(Y_u, A_u)\, du\right] \tag{4.8}$$

for every positive Borel function F on $\mathbb{R}_+ \times \mathbb{R}_+$. In particular for specifying the potential kernel U, it is sufficient to compute (4.8) for

$$F(y,t) = e^{-\lambda t} f(y)$$

with $\lambda \geqq 0$ and f bounded Borel. In this case, we denote the potential (4.8) by $U_\lambda f(x)$, that is

$$U_\lambda f(x) = E_x[\int_0^\infty e^{-\lambda A_u} f(Y_u)\, du], \tag{4.9}$$

(4.10) THEOREM. Let $\lambda \geqq 0$ be fixed. Then, U_λ is the potential operator for a Markov process $(\hat{Y}_u)$. The domain of the corresponding extended generator G_λ includes all bounded differentiable functions φ on $\mathbb{R}_+$. For such a function φ, denoting its derivative by φ',

$$G_\lambda \varphi(x) = -\lambda \varphi(x)\, a(x) + \varphi'(x)\, b(x) - \varphi(x) \int_0^\infty \mu(dw)\,(1 - e^{-\lambda g(y,w)})$$
$$+ \int_0^\infty \frac{dz}{z^2} [\varphi(y+k(y,x)) \int_0^1 e^{-\lambda h(y,z,w)}\, dw - \varphi(y)], \tag{4.11}$$

where a, b, g, h, k are the parameters appearing in (3.2) and (4.1) and where μ is related to the Poisson random measure M appearing in (4.1) so that the mean measure of M is $ds\, \mu(dw)$.

Moreover, the operator U_λ is the inverse of $-G_\lambda$, that is, $U_\lambda f = \varphi$ if and only if $-G_\lambda \varphi = f$.

REMARK. Note that $G_0 = G$, where G is the generator of Y described by (3.12).

PROOF. (a) Let F be a bounded differentiable function on $\mathbb{R}_+ \times \mathbb{R}_+$ and let F_1 and F_2 denote its partial derivatives with respect to the first and second arguments respectively. Then, by the change of variable formula used together with (3.12) and (4.1), we have

$$F(A_u, Y_u) = F(0, Y_0) + \int_0^u F_1(A_s, Y_s)\, a(Y_s)\, ds + \int_0^u F_2(A_s, Y_s)\, b(Y_s) ds$$
$$+ \sum_{0<s\leqq u} [F(A_s, Y_s) - F(A_{s-}, Y_{s-})]$$

$$= F(0,Y_0) + \int_0^u [F_1(A_s,Y_s)\, a(Y_s) + F_2(A_s,Y_s)\, b(Y_s)]\, ds$$

$$+ \int_0^u \int_0^\infty [F(A_{s-}+g(Y_{s-},w),Y_{s-}) - F(A_{s-},Y_{s-})]\, M(ds,dw)$$

$$+ \int_0^u \int_0^\infty \int_0^1 [F(A_{s-}+h(Y_{s-},z,w),Y_{s-}+k(Y_{s-},z))$$

$$- F(A_{s-},Y_{s-})]\, \hat{N}(ds,dz,dw).$$

Note that the integrands in the last two terms are measurable functions of the left-continuous process (Y_{s-},A_{s-}), and thus, predictable. Hence, the expectation of the last two terms will be the expectation of the same integrals but with the Poisson random measures replaced by their mean measures. Moreover, in the resulting integrals with respect to the Lebesgue measure ds, we may replace Y_{s-} by Y_s and A_{s-} by A_s since such alteration affect only countably many s and the Lebesgue measure of all such s is zero. Thus,

$$E_x[F(A_u,Y_u)] = F(0,x) + E_x\{\int_0^u [F_1(A_s,Y_s)\, a(Y_s) + F_2(A_s,Y_s)\, b(Y_s)]\, ds$$

$$+ \int_0^u ds \int_0^\infty \mu(dw)\, [F(A_s+g(Y_s,w),Y_s) - F(A_s,Y_s)]$$

$$+ \int_0^u ds \int_0^\infty \frac{dz}{z^2} \int_0^1 dw\, [F(A_s+h(Y_s,z,w),Y_s+k(Y_s,z)) - F(A_s,Y_s)]\}.$$

(b) In the preceding formula, let $F(t,y) = e^{-\lambda t}\varphi(y)$ for some bounded differentiable function φ with derivative φ'. Note that, with $G_\lambda\varphi$ defined by (4.11), we obtain

$$E_x[e^{-\lambda A_u}\varphi(Y_u)] = \varphi(x) + \int_0^u E_x[e^{-\lambda A_s} G_\lambda\varphi(Y_s)]\, ds. \tag{4.12}$$

Defining

$$\hat{P}_t\varphi(x) = E_x[e^{-\lambda A_t}\varphi(Y_t)], \tag{4.13}$$

the formula (4.12) becomes

$$\hat{P}_u \varphi(x) = \varphi(x) + \int_0^u \hat{P}_s (G_\lambda \varphi)(x)\, ds. \tag{4.14}$$

(c) Let $\hat{M}_t = E_x[\exp(-\lambda A_t) \mid Y_s;\ s \geqq 0]$. It follows from the additivity of (A_u) that this conditional expectation defines a multiplicative functional $t \to \hat{M}_t$ of the process Y (see ÇINLAR [2] for this point and [1] for the definitions and the claims to follow). Note that (4.13) yields

$$\hat{P}_t \varphi(x) = E_x[\hat{M}_t\ \varphi(Y_t)], \quad t \geqq 0,$$

which shows that $(\hat{P}_t)_{t \geqq 0}$ is a possibly sub-Markovian transition semigroup. Then, it is well known that $(\hat{P}_t)$ is the transition semigroup of the Markov process $(\hat{Y}_t)$ obtained from (Y_u) by killing via the multiplicative functional $(\hat{M}_t)$. Now, (4.14) becomes the integrated form of Kolmogorov's forward equations, and in particular, (4.14) shows that G_λ is the infinitesimal generator (at least in the extended sense of Dynkin) corresponding to the semigroup $(\hat{P}_t)$ of the process $(\hat{Y}_t)$.

(d) Since Y is increasing to $+\infty$ by assumption, Y is transient. Therefore, the killed process $\hat{Y}$ is transient even when $\lambda = 0$. Thus, the potential

$$\begin{aligned} U_\lambda \varphi(x) &= E_x[\int_0^\infty e^{-\lambda A_u}\ \varphi(Y_u)\ du] \\ &= E_x[\int_0^\infty \hat{M}_u\ \varphi(Y_u)\ du] = \int_0^\infty \hat{P}_u \varphi(x)\ du \end{aligned}$$

is finite for φ bounded. Note that, this last formula identifies U_λ as the potential operator corresponding to the process $(\hat{Y}_t)$. Finally, it is well-known that the potential and the negative of the infinitesimal generator are inverses of each other, so that $U_\lambda = (-G_\lambda)^{-1}$. □

5. SEMIMARKOV PROCESSES

Let (Y_u, A_u) be the Markov additive process described in the preceding section, Theorem (4.2). Define

$$C_t = \inf\{u: A_u > t\}, \quad t \geqq 0, \tag{5.1}$$

that is, the right-continuous functional inverse of (A_u). We interpret A_u as the actual time when an intrinsic clock shows u and C_t as the clock time when the actual time is t. Interpreting Y_u as the deterioration level when the clock time is u, then, the deterioration process in actual time is

$$X_t = Y_{C_t}, \quad t \geqq 0. \tag{5.2}$$

The processes X and Y are related to each other just as they were in Section 3; see Figure 5. However, because of the much more general nature of (A_u) here, the process X *is no longer Markovian.*

(5.3) THEOREM. Let (Y_u) be an Ito process satisfying (3.2), let (A_u) be defined by (4.1) and satisfy the conditions (4.3), and define (X_t) through (5.1), (5.2). Then, (X_t) is an increasing right-continuous semimarkov process in the sense of Definition (2.9).

PROOF. Let $\mathbb{M}$ be as defined by (2.8) for the present process X, and let T be a stopping time of X such that $T(\omega) \in \mathbb{M}(\omega)$ for almost every ω in $\{T<\infty\}$. Note that (see Figure 5 below), for almost every $\omega \in \{T<\infty\}$, $T(\omega)$ is a value attained by the function $u \to A_u(\omega)$ and moreover $T(\omega)$ does not belong to any interval $[A_{u-}(\omega), A_u(\omega))$, $u \geqq 0$. It follows that

$$A_{C_T} = T \quad \text{a.s. on} \quad \{T < \infty\},$$

which implies that any event determined by the future $\{X_{T+t};\ t \geqq 0\}$ is an event determined by $\{(Y_{U+u}, A_{U+u} - A_U);\ u \geqq 0\}$ with $U = C_T$. Thus, the strong Markov property of (Y,A) applied at the stopping time $U = C_T$ implies the strong Markov property of X at such T. □

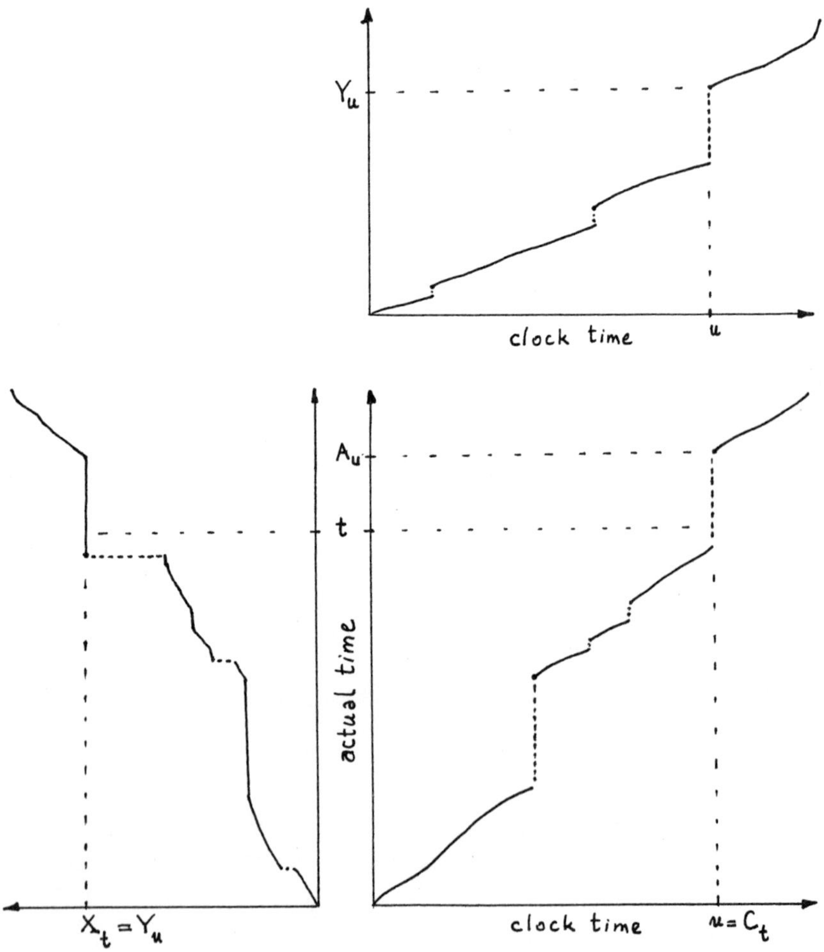

FIGURE 5. The semimarkov process X is obtained from the Markov additive process (Y,A) by a random time change using A as the actual time.

Note that the jumps of (A_u) correspond to intervals of constancy for (C_t) and, therefore, for (X_t). We have already exploited this fact while interpreting the structure of (A_u) in Section 4. In the converse direction, an interval of constancy for (A_u) corresponds to a jump of (C_t), that is, a jump in clock time at an instant of the actual time. This feature seems useless in applications in reliability, and can be avoided by making sure that $u \to A_u$ is strictly increasing.

Suppose $u \to A_u$ is strictly increasing. Then, $t \to C_t$ is continuous, but generally has intervals of constancy. If, further, $u \to A_u$ is continuous, then A_u is equal to the first term of the right side of (4.2), $t \to C_t$ is continuous and strictly increasing, and the process (X_t) reduces to a Hunt process as in Section 3.

Suppose $u \to A_u$ is strictly increasing as before. Recall that the jumps of it are intervals of constancy for both (C_t) and (X_t). During such intervals, (X_t) does not enjoy the strong Markov property. Such intervals are called contiguous to the set $\mathbb{M}$ defined in (2.8). In fact, we have

$$\mathbb{M}(\omega) = \{t\colon X_{t+\varepsilon}(\omega) > X_t(\omega) \quad \text{for every} \quad \varepsilon > 0\}$$

$$\subset \{t\colon A_u(\omega) = t \quad \text{for some} \quad u \geqq 0\},$$

that is, $\mathbb{M}(\omega)$ is contained in the range of the function $u \to A_u(\omega)$. Further, the complement $\mathbb{R}_+ \setminus \mathbb{M}(\omega)$ of the set $\mathbb{M}(\omega)$ is a countable union of intervals of type $[\ )$, and those intervals are the contiguous intervals.

Let $T_1, T_2, \ldots$ be some stopping times of the process X and suppose that the T_n are strictly increasing to a limit T. Then, T is again a stopping time, and is said to be *predictable* (since it is foretold by the T_n). Consider now the stopping times $C_{T_1}, C_{T_2}, \ldots, C_T$ of the Markov additive process (Y_u, A_u).

It follows from the strict monotonicity of $u \to A_u$ that $t \to C_t$ is continuous and hence

$$\lim_n C_{T_n} = C_T.$$

If, further, the stopping time T is the left-end-point of a contiguous interval, then C_T is the time of the jump of $u \to A_u$ corresponding to that contiguous interval, and it is possible to find stopping times U_n of (Y_u, A_u) that are strictly increasing to C_T. Hence, if a stopping time T of (X_t) is the left-end-point of a contiguous interval and is predictable, then C_T is a predictable stopping time of (Y_u, A_u) and is a time of jump. But, this is impossible: jumps of (Y_u, A_u) are governed by Poisson random measures, and Poisson random measures do not have atoms at predictable times. It follows that we have the following limitation on the semimarkov process (X_t).

(5.5) PROPOSITION. Let (X_t) be the semimarkov process described in Theorem (5.3), suppose that $u \to A_u$ is strictly increasing, and let T be a stopping time of (X_t) that belongs to the set of left-end-points of the contiguous intervals of $\mathbb{M}$. Then, T is totally unpredictable, that is, the probability that T equals S is zero for every predictable stopping time S of (X_t). □

The following shows that, subject to the properties discussed above (and in the preceding proposition), every increasing semimarkov process can be obtained in this manner. In other words, the construction above can be reversed.

(5.6) THEOREM. Let (X_t) be an increasing right-continuous semimarkov process in the sense of Definition (2.9). Suppose that $\mathbb{M}$ defined by (2.8) has no isolated points, and every stopping time belonging to the set of left-end-points of the contiguous intervals of $\mathbb{M}$ is totally unpredictable. Then, there exists a Markov additive process (Y,A) such that Y is an Ito process satisfying (3.2), A has the form (4.1) and is strictly increasing, and X is obtained from (Y,A) through (5.1) - (5.2).

PROOF. Let X be as specified. Then, JACOD [10] has shown the existence of a quasi-left-continuous strong Markov additive process $(\hat{Y},\hat{A})$ such that A is strictly increasing and (we are writing $Y(t)$ for Y_t etc.)

$$X_t = \hat{Y}(\hat{C}_t), \quad \hat{C}_t = \inf\{s: \hat{A}_s > t\}. \tag{5.7}$$

In general, the Markov additive process $(\hat{Y},\hat{A})$ does not satisfy (3.2) and (4.1). Applying the basic representation theorem of ÇINLAR and JACOD [6] to the additive semimartingale $(\hat{Y}_s-\hat{Y}_0,\hat{A}_s)$, we arrive at the existence of a Markov additive process (Y_u,A_u) satisfying (3.2) and (4.1) and such that

$$\hat{Y}_s = Y(\check{C}_s), \quad \hat{A}_s = A(\check{C}_s) \tag{5.8}$$

where

$$\check{C}_s = \inf\{u: \check{A}_u > s\} \tag{5.9}$$

for some strictly increasing and continuous additive functional $\check{A}$ of (Y_u). Then, $\check{C}$ is also strictly increasing and continuous, and we have

$$A_u = \hat{A}(\check{A}_u), \tag{5.10}$$

which shows that A is strictly increasing (since $\hat{A}$ is so).

There remains to show that the effect of two successive time changes can be produced by one time change. It follows from (5.7) and (5.8) that

$$X_t = \hat{Y}(\hat{C}_t) = Y(\check{C}(\hat{C}_t)), \tag{5.11}$$

and by (5.9),

$$\begin{aligned} \check{C}(\hat{C}_t) &= \inf\{u: \check{A}_u > \hat{C}_t\} \\ &= \inf\{u: \check{A}_u > \inf\{s: \hat{A}_s > t\}\} \\ &= \inf\{u: \hat{A}(\check{A}_u) > t\} \\ &= \inf\{u: A_u > t\} \end{aligned} \tag{5.12}$$

in view of (5.10). Thus, putting

$$C_t = \check{C}(\hat{C}_t) \tag{5.13}$$

into (5.11) and (5.12) shows that X is related to (Y,A) as claimed. □

It is instructive to compare the characterization here with that given for the continuous semimarkov processes in Section 2. For X continuous, Y becomes trite, $Y_u = Y_0 + u$, and A has independent increments. However, in this case, A here is slightly less general than that given in (2.11) in that we no longer have jumps for A_u at fixed u_i.

6. LIFETIME AND THE DETERIORATION LEVEL AT FAILURE

Throughout this section we assume that the deterioration process (X_t) is an increasing right-continuous semimarkov process obtained from a Markov additive process (Y_u,A_u) as described in Theorem (5.3) and further assume that (A_u) is finite valued, strictly increasing, and with

$\lim_{u\to\infty} A_u = \infty$. As we have seen in Theorem (5.6), this covers most semimarkov processes (X_t).

Suppose that the failure occurs when X exceeds a fixed threshold value ξ (see Remark (6.11) for the generalization to random threshold values). Then

$$T = \inf\{t: X_t > \xi\}, \qquad V = X_T \tag{6.1}$$

are the lifetime and the deterioration level at the time of failure respectively. Generally, $V \geqq \xi$ and cost of repair or replacement may depend on V. Thus, it is useful to have the joint distribution of T and V, assuming that $X_0 = x < \xi$.

The following relates the lifetime T and the failure level V to the process (Y_u, A_u).

(6.2) PROPOSITION. Let

$$U = \inf\{u: Y_u > \xi\}. \tag{6.3}$$

Then,

$$U = C_T, \qquad Y_U = X_T = V, \tag{6.4}$$

$$T = \begin{cases} A_{U-} & \text{on} \quad \{Y_U > \xi\}, \\ \\ A_U & \text{on} \quad \{Y_U = \xi\}. \end{cases} \tag{6.5}$$

PROOF. Since A is strictly increasing, its functional inverse C is continuous. Thus, as t varies, C_t takes on all the values in $\mathbb{R}_+ = [0,\infty)$, and therefore, $X_t = Y(C_t)$ takes on all the values that $u \to Y_u$ does. This shows, together with an examination of Figure 5, that (6.4) holds with U defined by (6.3).

Fix $\omega \in \Omega$ and suppose $T(\omega) = t$, $X_t(\omega) = V(\omega) = \xi$, $U(\omega) =$

$C_T(\omega) = u$. Then, $X_{t+\varepsilon}(\omega) > \xi$ for every $\varepsilon > 0$ by the definition of $T(\omega)$, and $A_u(\omega) \geqq t$ by the definition of $C_t(\omega)$. But $A_u(\omega) > t$ would imply that $X_{t+\varepsilon}(\omega) = X_t(\omega) = \xi$ for $\varepsilon \leqq A_u(\omega) - t$. Hence, we must have $A_u(\omega) = t$ as claimed in (6.5) for this case $Y_u(\omega) = \xi$.

Fix $\omega \in \Omega$ again and suppose $T(\omega) = t$, $U(\omega) = u$, $X_t(\omega) = Y_u(\omega) = x > \xi$. If $A_{u-}(\omega) < t$, since $A_u(\omega) \geqq t$ by the definition of $C_t(\omega) = u$, we would have $X_s(\omega) = X_t(\omega) = x > \xi$ for all $s \in [A_{u-}(\omega), A_u(\omega)) \supset [A_{u-}(\omega), t)$. Since this contradicts the definition of $T(\omega)$, we must have $A_u(\omega) = t$ as claimed by (6.5) in this case where $Y_u(\omega) > \xi$.

The following is immediate from (6.4) - (6.5) and is the starting point of computations.

(6.6) COROLLARY. For any positive Borel function F on $\mathbb{R}_+ \times \mathbb{R}_+$,

$$F(T,V) = F(A_U,\xi) + [F(A_{U-},Y_U) - F(A_U,\xi)]\, I_{\{Y_U > \xi\}}. \tag{6.7}$$

To find the joint distribution of (T,V), it is sufficient to compute $E_x[F(T,V)]$ for every positive Borel function F on $\mathbb{R}_+ \times \mathbb{R}_+$, and by standard results on approximating such functions, we may further limit ourselves to computing $E_x[F(T,V)]$ for positive Borel functions F on $\mathbb{R}_+ \times \mathbb{R}_+$ which are differentiable in the first argument. In fact, we could limit ourselves to $F(t,v) = e^{-\alpha t} f(v)$, but the computations do not simplify enough to warrant such restrictions.

(6.8) THEOREM. Let F be a bounded Borel function on $\mathbb{R}_+ \times \mathbb{R}_+$ and differentiable in its first argument (write F_1 for its partial derivative). Then

$$E_x[F(T,V)] = F(0,\xi) + \int_0^\xi \int_0^\infty U(x,dy,dt)\, \hat{F}(y,t) \tag{6.9}$$

where $U(x,dy,dt)$ is the potential kernel defined by (4.8) and

$$\hat{F}(y,t) = F_1(t,\xi)\, a(y) + \int_0^\infty \mu(dw)\, [F(t+g(y,w),\xi) - F(t,\xi)]$$
$$+ \int_0^\infty \frac{dz}{z^2} \int_0^1 dw\, [F(t,y+k(y,z))\, 1_{(\xi,\infty)}(y+k(y,z))$$
$$+ F(t+h(y,z,w),\xi)\, 1_{[0,\xi]}(y+k(y,z)) - F(t,\xi)]. \tag{6.10}$$

(6.11) REMARK. Instead of being fixed, threshold value may be a random variable ξ independent of the deterioration process and with some distribution γ. Then, it is easy to see from (6.9) that

$$E_x[F(T,V)] = \int_0^\infty \gamma(dz)\, [F(0,z) + \int_0^z \int_0^\infty U(x,dy,dt)\, \hat{F}(y,t)]. \quad \square$$

We shall limit ourselves to fixed ξ from here on.

(6.12) COROLLARY. Let φ be a bounded Borel function on $\mathbb{R}_+$ and $\lambda > 0$. Then

$$E_x[e^{-\lambda T}\, \varphi(V)] = \varphi(\xi) + U_\lambda f(x)$$

where U_λ is as defined by (4.9) and discussed in Theorem (4.10) and where $f(y) = 0$ for $y > \xi$ and we have, for $y \leq \xi$,

$$f(y) = -\lambda\varphi(\xi)\, a(y) - \varphi(\xi) \int_0^\infty \mu(dw)\, (1 - e^{-\lambda g(y,w)})$$
$$+ \int_0^\infty \frac{dz}{z^2} \int_0^1 dw\, [\varphi(y+k(y,z))\, 1_{(\xi,\infty)}(y+k(y,z))$$
$$+ e^{-\lambda h(y,z,w)}\, \varphi(\xi)\, 1_{[0,\xi]}(y+k(y,z)) - \varphi(\xi)]. \quad \square$$

PROOF of Corollary is immediate from Theorem (6.8) where we take $F(t,y) = e^{-\lambda t}\, \varphi(y)$. A further corollary of some interest is that on the marginal distribution of T. This can be obtained from Theorem (6.8) by taking $F(t,y) = f(t)$. The result is as follows for f differentiable and bounded:

$$E_x[f(T)] = f(0) + \int_0^{\xi}\int_0^{\infty} U(x,dy,dt)\,\hat{F}(y,t) \tag{6.13}$$

where

$$\begin{aligned}\hat{F}(y,t) &= f'(t)\,a(y) + \int_0^{\infty} \mu(dw)\,[f(t+g(y,w)) - f(t)]\\ &\quad + \int_0^{\infty} \frac{dz}{z^2} \int_0^1 dw\,[f(t+h(y,z,w)) - f(t)]\,1_{[0,\xi]}(y+k(y,z)).\end{aligned} \tag{6.14}$$

Finally, the Laplace transform of T can be obtained by substituting $\varphi(x) = 1$ into (6.12). This yields

$$E_x[e^{-\lambda T}] = 1 - U_\lambda f(x) \tag{6.15}$$

where $f(y) = 0$ for $y > \xi$ and, for $y \leq \xi$,

$$\begin{aligned}f(y) &= +\lambda a(y) + \int_0^{\infty} \mu(dw)\,[1 - e^{-\lambda g(y,w)}]\\ &\quad + \int_0^{\infty} \frac{dz}{z^2} \int_0^1 dw\,[1 - e^{-\lambda h(y,z,w)}]\,1_{[0,\xi]}(y+k(y,z)).\end{aligned} \tag{6.16}$$

For computing the mean of T, we can use (6.13) with $f(t) = t$ and similarly for the higher moments.

PROOF OF THEOREM (6.8). We start with the expression (6.7), derive expressions for each term on the right side, and take expectations.

(a) Consider the term $\alpha = F(A_U,\xi)$. Proceeding just as in part (a) of the proof of Theorem (4.10) we have

$$\begin{aligned}\alpha &\equiv F(A_U,\xi)\\ &= F(0,\xi) + \int_0^{U} F_1(A_s,\xi)\,a(Y_s)\,ds + \sum_{s\leq U} [F(A_s,\xi) - F(A_{s-},\xi)]\end{aligned} \tag{6.17}$$

$$= F(0,\xi) + \int_0^U F_1(A_s,\xi)\, a(Y_s)\, ds$$

$$+ \int_0^U \int_0^\infty [F(A_{s-}+g(Y_{s-},w),\xi) - F(A_{s-},\xi)]\, M(ds,dw)$$

$$+ \int_0^U \int_0^\infty \int_0^1 [F(A_{s-}+h(Y_{s-},z,w),\xi) - F(A_{s-},\xi)]\, \hat{N}(ds,dz,dw).$$

Note that we can replace the integrals over $[0,U]$ by integrals over $[0,\infty]$ with the term $1_{[0,U]}(s) = 1_{[0,\xi]}(Y_{s-})$ put into the integrands. Then, taking expectations as in the proof of (4.10) still, we get

$$E_x[\alpha] = F(0,\xi) + E_x\{\int_0^\infty F_1(A_s,\xi)\, a(Y_s) 1_{[0,\xi]}(Y_s)\, ds$$

$$+ \int_0^\infty \int_0^\infty [F(A_s+g(Y_s,w),\xi) - F(A_s,\xi)]\, 1_{[0,\xi]}(Y_s)\, ds\, \mu(dw)$$

$$+ \int_0^\infty \int_0^\infty \int_0^1 [F(A_s+h(Y_s,z,w),\xi) - F(A_s,\xi)]\, 1_{[0,\xi]}(Y_s)\, ds \cdot \frac{dz}{z^2}\, dw\}$$

$$= F(0,\xi) + \int_0^\xi \int_0^\infty U(x,dy,dt)\, \overline{F}(y,t)$$

where

$$\overline{F}(y,t) = F_1(t,\xi)\, a(y) + \int_0^\infty \mu(dw)\, [F(t+g(y,w),\xi) - F(t,\xi)]$$

$$+ \int_0^\infty \frac{dz}{z^2} \int_0^1 dw\, [F(t+h(y,z,w),\xi) - F(t,\xi)]. \qquad (6.19)$$

(b) Consider the second term on the right side of (6.7). Note that

$$\beta \equiv [F(A_{U-},Y_U) - F(A_U,\xi)]\, I_{\{Y_U>\xi\}}$$

$$= \sum_s [F(A_{s-},Y_s) - F(A_s,\xi)]\, I_{\{Y_{s-}\le\xi,\ Y_s>\xi\}}, \qquad (6.20)$$

the sum having at most one term, namely for $s = U$, since the event $\{Y_{s-} \leq \xi, Y_s > \xi\}$ is empty for all other s. Now using the constructions (3.2) and (4.1) of Y and A, we get

$$\beta = \int_0^\infty \int_0^\infty \int_0^1 [F(A_{s-}, Y_{s-}+k(Y_{s-},z)) - F(A_{s-}+h(Y_{s-},z,w),\xi)]$$

$$I_{\{Y_{s-} \leq \xi,\ Y_{s-}+k(Y_{s-},z)>\xi\}} \hat{N}(ds,dz,dw). \tag{6.21}$$

Taking expectations, via the arguments of the proof of (4.10), gives

$$E_x[\beta] = E_x \int_0^\infty \int_0^\infty \int_0^1 [F(A_s,Y_s+k(Y_s,z)) - F(A_s+h(Y_s,z,w),\xi)]$$

$$1_{[0,\xi]}(Y_s)\ 1_{(\xi,\infty)}(Y_s+k(Y_s,z))\ ds \cdot \frac{dz}{z^2}\ dw\}$$

$$= \int_0^\xi \int_0^\infty U(x,dy,dt)\ \bar{\bar{F}}(y,t) \tag{6.22}$$

where

$$\bar{\bar{F}}(y,t) = \int_0^\infty \frac{dz}{z^2} \int_0^1 dw\ [F(t,y+k(y,z)) - F(t+h(y,z,w),\xi]\ 1_{(\xi,\infty)}(y+k(y,z)). \tag{6.23}$$

The proof of Theorem (6.8) follows now from (6.18) and (6.22), by adding them side by side and putting $\hat{F}(y,t) = \bar{F}(y,t) + \bar{\bar{F}}(y,t)$, which yields (6.10) from (6.19) and (6.23). □

REFERENCES

1. R.M. BLUMENTHAL and R.K. GETOOR. *Markov Processes and Potential Theory*, Academic Press, New York, 1968.

2. E. ÇINLAR. Markov additive processes, II. *Z. Wahrscheinlichkeitstheorie verw. Geb.* *24* (1972), 94 - 121.

3. E. ÇINLAR. Lévy systems of Markov additive process. *Z. Wahrscheinlichkeitstheorie verw. Geb. 31* (1975), 175 - 185.

4. E. ÇINLAR. Shock and wear models and Markov additive processes. In *The Theory and Applications of Reliability, 1*, ed. by I.N. Shimi and C.P. Tsokos. Academic Press, New York, 1977.

5. E. ÇINLAR. On increasing continuous processes. *Stochastic Processes and Their Applications 9* (1979), 147 - 154.

6. E. ÇINLAR and J. JACOD. Representation of semimartingale Markov processes in terms of Wiener processes and Poisson random measures. *Seminar on Stochastic Processes 1981*, pp. 159 - 242. Birkhäuser, Boston, 1981.

7. J.W. DROSEN. Failure Times and Optimal Stopping Rules of Generalized Shock Models. Unpublished dissertation, Northwestern University, 1983.

8. J.D. ESARY, A.W. MARSHALL, and F. PROSCHAN. Shock models and wear processes. *Ann. Probab. 1* (1973), 627 - 649.

9. K. ITO. *Stochastic Processes*. Lecture Notes Series No. 16, Matematisk Institut, Aarhus Universitet, 1969.

10. J. JACOD. Systèmes régénératifs et processus semi-markoviens, *Z. Wahrscheinlichkeitstheorie verw. Geb. 31* (1974), 1 - 23.

11. P. LÉVY. Processus semi-Markoviens. *Proc. Int. Congr. Math. (Amsterdam) 3* (1954), 416 - 426.

WEAR AND DAMAGE PROCESSES FROM SHOCK MODELS IN RELIABILITY THEORY

Moshe Shaked [1]

Department of Mathematics
University of Arizona
Tucson, Arizona

In many applications the time to failure of a device or a system may be represented as a first passage time for an appropriate stochastic process. In this survey paper we consider situations in which the appropriate stochastic process is a damage or wear process. The time to failure then is the first time the damage reaches or crosses the breaking threshold of the device. Various ways of modeling accumulation of wear or damage using shock models are discussed. Processes with these kinds of first passage times, being IFR, IFRA, NBU, NBUE, HNBUE or having logconcave densities, are given. Multivariate analogs of these processes are also discussed.

[1] *Supported by NSF Grant MCS 82-00098.*

ISBN 0-12-041420-1

1. INTRODUCTION AND CONVENTIONS

In many applications the time to failure of a device or of a system may be represented as a first passage time for an appropriate stochastic process. A number of such waiting times have logconcave densities or increasing failure rate (IFR), increasing failure rate average (IFRA), new better than used (NBU), new better than used in expectation (NBUE) or harmonic new better than used in expectation (HNBUE) distributions.

In this survey paper we consider situations in which the appropriate stochastic process $\{Z(t), t \geq 0\}$ is a damage process, that is $Z(t)$ is the damage (or wear) level of the device at time t. The time to failure of the device then is the first time t at which $Z(t) \geq x$, where x is the breaking threshold of the device.

There are various ways of modeling accumulation of damage or wear. In this paper we restrict our discussion to shock models. In these models, shocks occur randomly in time and random damages are added to the current damage level of the device at the times of these shocks. Furthermore, in some of these models, the damage level may decrease between shocks due to repairs or self-recovery procedures. Duals of these kinds of processes and multivariate analogs of these are also discussed.

The gamma process can be shown to be a limit of sequences of some of the damage processes discussed in this paper. Although we do not discuss this process in detail in this paper, we refer the interested reader to Abdel-Hameed (1975). He showed that, under some assumptions on the intensity function, the first passage times of the Gamma process to upper sets are IFR, IFRA, or NBU. Block and Savits (1980a), Section 5, further studied the Gamma process.

In some situations, the waiting time for the failure of a device can be represented as a first passage time for a stochastic process which cannot be derived from the shock models discussed in this paper. For example, first passage times to upper sets of various Markov and semi-Markov processes have logconcave densities or are IFR, IFRA or NBU. These processes are not discussed in this paper (although some of the processes discussed in this paper are Markovian) but the reader is referred to Marshall and Shaked (1983a) and references therein.

Conventions. In this paper the terms increasing and decreasing are not used in the strict sense. The adverb 'almost surely' is often omitted. For example, we say that a stochastic process is increasing instead of saying that it is almost surely nondecreasing. All the stochastic processes which are considered in the paper are assumed to be right-continuous even if we do not say it explicitly. Similarly, unless stated otherwise, all the functions mentioned in the paper are assumed to be measurable.

2. DEFINITIONS

Every distribution of a nonnegative random variable is called a life distribution. A random variable or its distribution F or its survival function $\bar{F} = 1 - F$ is said to be or to have:

(i) a PF_2 density if F has a density which is logconcave;

(ii) increasing failure rate (IFR) if $\bar{F}$ is logconcave;

(iii) increasing failure rate average (IFRA) if $-\log \bar{F}$ is starshaped, that is, if $-t^{-1} \log \bar{F}(t)$ is increasing in $t \geq 0$;

(iv) new better than used (NBU) if $-\log \bar{F}$ is superadditive (that is, if $\bar{F}(x + t) \leq \bar{F}(x)\,\bar{F}(t)$ for all $x \geq 0$, $t \geq 0$);

(v) new better than used in expectation (NBUE)

if $\infty > \int_0^\infty \overline{F}(x)dx \geq \int_0^\infty (\overline{F}(t+x)/\overline{F}(t))dx$ for all $t \geq 0$;

(vi) harmonic new better than used in expectation (HNBUE) if $\infty > \int_0^\infty \overline{F}(x)dx \equiv \mu$ and $\int_t^\infty \overline{F}(x)dx \leq \mu \exp\{-t/\mu\}$ for all $t \geq 0$.

Note that every random variable which is degenerate on some positive value is IFR, IFRA, NBU, NBUE and HNBUE.

The following implications are well known (see, e.g., Barlow and Proschan (1975) and Klefsjo (1981b)):

$$PF_2 \text{ density} \Rightarrow \text{IFR} \Rightarrow \text{IFRA} \Rightarrow \text{NBU} \Rightarrow \text{NBUE} \Rightarrow \text{HNBUE}. \tag{2.1}$$

A nonnegative stochastic process $\{X(t), t \geq 0\}$ is called a PF_2 [respectively, IFR, IFRA, NBU, NBUE, HNBUE] process if $T_x \equiv \inf\{t : X(t) \geq x\}$ is PF_2 [respectively, IFR, IFRA, NBU, NBUE, HNBUE] for every $x \geq 0$. From (2.1) it follows, for example, that every PF_2 process is also an IFR, IFRA, NBU, NBUE and HNBUE process. Similar facts follow from the other implications of (2.1).

3. A UNIVARIATE CUMULATIVE DAMAGE SHOCK MODEL

Esary, Marshall and Proschan (1973) studied the following cumulative damage shock model:

Model 3.1 (i) A device is subjected to shocks occurring randomly in time as events in a Poisson process.

(ii) At time $t = 0$ the damage level of the device is 0. Upon occurrence of the ith shock, the device suffers a nonnegative random damage with distribution F_i. The damages are independent of the process in (i).

(iii) Each random damage, at the time of its occurrence, adds to the current damage level of the device.

(iv) Between shocks the damage level stays constant.

Let $\{N(t), t \geqslant 0\}$ denote the Poisson process of Model 3.1(i) and let $X_1, X_2, \ldots$ be the random damages, so that X_i is distributed according to F_i, $i = 1,2,\ldots$. At the moment we do not assume that the X_i's are independent. Let $S_0 \equiv 0$, $S_i = X_1 + X_2 + \ldots + X_i$, $i = 1,2,\ldots$. The process $\{Z(t) \equiv S_{N(t)}, t \geqslant 0\}$ is called the damage process associated with Model 3.1.

In Sections 4-6 it is shown, under various assumptions, that the damage process associated with Model 3.1 is PF_2, IFR or IFRA process. Also, other cumulative damage shock and wear models, whoch give rise to such processes, will be described.

4. PF_2 PROCESSES

Esary, Marshall and Proschan (1973) have found conditions under which the damage process Z associated with Model 3.1 is PF_2:

Theorem 4.1. If in Model 3.1 (ii)

(i) $X_1, X_2, \ldots$ are independent;

(ii) $F_i = F$, $i = 1,2,\ldots$, for some F;

(iii) F has a PF_2 density;

then Z is a PF_2 process.

Assaf, Shaked and Shanthikumar (1983) have recently obtained various extensions of this result.

The conclusion of Theorem 4.1 says that the first passage time of Z to x has a PF2 density for every $x \geqslant 0$. It does not necessarily follow then that if the threshold x is random then $\inf\{ t:Z(t) \geq x \}$ has a PF_2 density. Thus, the following result of Esary, Marshall and Proschan (1973) is of interest:

Theorem 4.2. For Model 3.1, if (i), (ii) and (iii) of Theorem 4.1 hold, if x is a random variable independent of X_1, $X_2, \ldots$ and of the process of Model 3.1(i) and if x has a PF_2 density then $\inf\{ t:Z(t) \geq x \}$ has a PF_2 density.

5. IFR PROCESSES

Esary, Marshall and Proschan (1973) obtained conditions under which the damage process Z associated with Model 3.1 is IFR:

Theorem 5.1. If in Model 3.1(ii)

(i) $X_1, X_2, \ldots$ are independent;

(ii) $F_i = F$, $i = 1,2,\ldots,$ for some F;

(iii) F is logconcave;

then Z is an IFR process.

Generalizations of Theorem 5.1 can be obtained from a combination of Theorem 4.9 of Esary, Marshall and Proschan (1973) with Theorem 2.4 of A-Hameed and Proschan (1975) or with Theorem 3.5 of Gottlieb (1980) or with Theorem 1 of Ghosh and Ebrahimi (1982). We omit the details.

In analogy with Theorem 4.2, Esary, Marshall and Proschan (1973) showed:

Theorem 5.2. For Model 3.1, if (i), (ii) and (iii) of Theorem 5.1 hold, if x is a random variable independent of X_1, $X_2, \ldots$ and of the process of Model 3.1(i) and if x has a PF_2 density then $\inf\{ t:Z(t) \geq x \}$ is IFR.

See also Abdel-Hameed (1983).

A-Hameed and Proschan (1973) considered the following

cumulative damage shock model:

Model 5.3. (i) A device is subjected to shocks occurring randomly in time as events in a nonhomogeneous Poisson process with intensity function $\lambda(t)$, $t \geq 0$.

(ii) At time $t = 0$ the damage level of the device is 0. Upon occurrence of the ith shock the device suffers a random damage which is a gamma variable with density $f_i(u) = [\Gamma(a_i)]^{-1} b^{a_i} u^{a_i - 1} e^{-bu}$, $u \geq 0$, $a_i > 0$, $b > 0$, $i = 1,2,\ldots$. The damages are independent of the process in (i).

(iii) Same as Model 3.1(iii).

(iv) Same as Model 3.1(iv).

Let $\{N(t), t \geq 0\}$ denote the nonhomogeneous Poisson process of Model 5.3(i), let $X_1, X_2, \ldots$ be the gamma random damages, let $S_0 \equiv 0$, $S_i = X_1 + X_2 + \ldots + X_i$, $i = 1,2,\ldots,$ and let $\{Z(t) = S_{N(t)}, t \geq 0\}$ be the associated damage process. Denote $A_0 \equiv 0$, $A_i = a_1 + \ldots + a_i$, $i = 1,2,\ldots,$ and

$$\Lambda(t) = \int_0^t \lambda(u)du, \quad t \geq 0.$$

A-Hameed and Proschan (1973) were interested in the case when the threshold x is random. However, a special case of one of their results (when x is nonrandom) is the following:

Theorem 5.4. If in Model 5.3,

(i) $X_1, X_2, \ldots$ are independent;

(ii) Λ is convex on $[0,\infty]$;

(iii) A_k, as a function of k, is convex on $0,1,\ldots$;

then the damage process Z associated with Model 5.3 is IFR.

6. IFRA PROCESSES

Model 3.1 also gives rise naturally to IFRA processes. Unlike the results of Sections 4-5, no assumptions on the form of F_i, $i = 1,2,\ldots,$ are needed. Esary, Marshall and Proschan (1973) proved:

Theorem 6.1. If in Model 3.1(ii)

(i) $X_1, X_2, \ldots$ are independent;

(ii) $F_i(u)$ is decreasing in i for all $u > 0$;

then Z is an IFRA process.

In particular:

Corollary 6.2. If in Model 3.1(ii)

(i) $X_1, X_2, \ldots$ are independent;

(ii) $F_i = F$, $i = 1,2,\ldots,$ for some F;

then Z is an IFRA process.

Esary, Marshall and Proschan (1973) also found conditions on the joint distribution of $X_1, X_2, \ldots$ under which Z is IFRA even if the X_i's are not independent. Furthermore, Esary, Marshall and Proschan (1973) showed, by example, that the assumptions of Corollary 6.2 are not sufficient to imply that Z is an IFR process. They also obtained analogs of Theorems 4.2 and 5.2 in the IFRA case.

Related results which, together with Lemma 4.1a of Esary, Marshall and Proschan (1973), indicate situations which give rise to IFRA processes are Theorem 2.6 of A-Hameed and Proschan (1975) and Theorems 2.1 and 2.3 of Klefsjo (1981a). See also Abdel-Hameed (1983).

A-Hameed and Proschan (1973) obtained the following analog of Theorem 5.4:

Theorem 6.3. If in Model 5.3

(i) $X_1, X_2, \ldots$ are independent;

(ii) $\Lambda(t)/t$ is increasing in $t \geq 0$;

(iii) A_k/k is increasing in $k = 1,2,\ldots$;

then the damage process Z associated with Model 5.3 is IFRA.

Ross (1979) essentially proved the following result which he called <u>the generalized IFRA closure theorem</u>:

<u>Theorem 6.4</u>. If $\{Z_i(t), t \geq 0\}$, $i = 1,2,\ldots, n$, are increasing independent IFRA processes then $\{\phi(Z_1(t),\ldots, Z_n(t)), t \geq 0\}$ is an IFRA process whenever ϕ is right-continuous and increasing.

Ross (1981) used Theorem 6.4 to obtain a useful generalization of Theorem 6.1.

7. NBU PROCESSES

A-Hameed and Proschan (1973) obtained an NBU analog of Theorem 6.3:

<u>Theorem 7.1</u>. If in Model 5.3

(i) $X_1, X_2,\ldots$ are independent;

(ii) $\Lambda(t + s) \geq \Lambda(t) + \Lambda(s)$, $t \geq 0$, $s \geq 0$;

(iii) $A_{k+\ell} \geq A_k + A_\ell$, $k=0,1,\ldots$, $\ell=0,1,\ldots$;

then the associated damage process Z is NBU.

Since the NBU property is weaker than the PF_2, IFR and IFRA properties, one can expect to find NBU processes in real situations which cannot be described by the restrictive assumptions of Models 3.1 and 5.3. The first step in relaxing the restrictions of Model 3.1 is by replacing the Poisson process in Model 3.1(i) by a more general point process. Consider the following model:

<u>Model 7.2</u>. (i) A device is subjected to shocks occurring randomly in time as renewal epochs in a renewal process.

(ii) Same as Model 3.1(ii).

(iii) Same as Model 3.1(iii).

(iv) Same as Model 3.1(iv).

Denote the interrenewal times of the renewal process in (i) by $B_1, B_2, \ldots$ and let Z be the associated damage process.

Combination of Lemma 4.1a of Esary, Marshall and Proschan (1973) and (2.1) of this paper with Theorem 2.7 of A-Hameed and Proschan (1975) or with Section 3 of Block and Savits (1978) gives:

Theorem 7.3. If in Model 7.2

(i) $B_1, B_2, \ldots$ are NBU;

(ii) $X_1, X_2, \ldots$ are independent;

(iii) $\bar{F}_i(u)$ is decreasing in i for all $u \geq 0$;

then the associated damage process is NBU.

See also Abdel-Hameed (1983).

Sumita and Shanthikumar (1983) relaxed the independence assumption in Model 7.2(ii). They considered the following model:

Model 7.4. (i) Same as Model 7.2(i).

(ii) At time $t = 0$ the damage level of the device is 0. Upon occurrence of the ith shock, the device suffers a nonnegative random damage with distribution F_i.

(iii) Same as Model 3.1(iii).

(iv) Same as Model 3.1(iv).

Denote the interrenewal times of the renewal process in (i) by $B_1, B_2, \ldots$, and denote the respective random damages by $X_1, X_2, \ldots$. Let Z by the associated damage process. Sumita and Shanthikumar (1983) proved:

Theorem 7.5. If in Model 7.4

(i) $B_1, B_2, \ldots$ are NBU

(ii) $(B_1, X_1), (B_2, X_2), \ldots$ are i.i.d.;

(iii) X_1 is stochastically increasing in B_1;
then the associated damage process Z is NBU.

El-Neweihi, Proschan and Sethuraman (1978) have essentially obtained an NBU analog of Theorem 6.4:

Theorem 7.6. If $\{Z_i(t), t \geq 0\}$, $i = 1,2,\ldots, n$, are increasing independent NBU processes then $\{\phi(Z_1(t), \ldots, Z_n(t)), t \geq 0\}$ is an NBU process whenever ϕ is right-continuous and increasing.

The processes which were discussed in Sections 4-6 and in this section up to this point were almost surely monotone (increasing). In practice one often encounters processes which need not be monotone. In the setting of Models 3.2, 5.3, 7.2 and 7.4 the associated damage process Z need not be monotone if, between shocks, the damage decreases in some deterministic fashion. Such recovery between shocks is typical to living organisms (see Marshall and Shaked (1983a) for more examples).

Marshall and Shaked (1983a) considered the following generalization of Model 3.1:

Model 7.7.
(i) Same as Model 7.2(i).
(ii) Same as Model 7.4(ii).
(iii) Same as Model 3.1(iii).
(iv) Between shocks the damage level decreases in some deterministic fashion (although the damage level never becomes negative).

Let $B_1, B_2, \ldots$ be the interenewal times in (i) and let $X_1, X_2, \ldots$ be the random damages in (ii). Denote the associated damage process by $\{Z(t), t \geq 0\}$.

In most applications the deterministic decrease of Z between shocks is of the form: "the rate of decrease at time t depends only on the level of damage at t provided t is not an epoch in the revewal process in (i)". This is the case, for example, if the decrease between times of shocks is linear or exponential.

Formally, such a decrease is determined by a strictly decreasing function $g:(0,\infty) \to (-\infty,\infty)$. Denote $R_0 \equiv 0$, $R_i = B_1 + \dots + B_i$, $i = 1,2, \dots$. The process Z, then, is defined by:

$$\begin{aligned} Z(t) &= 0 \quad \text{if} \quad 0 \leq t < R_1, &(7.1)\\ &= [g(t-R_i + g^{-1}(Z(R_i)))]^+ \\ &\qquad \text{if} \quad R_i < t < R_{i+1},\ i = 1,2,\dots, \\ &= Z(R_i-) + X_i \\ &\qquad \text{if} \quad t = R_i,\ i = 1,2,\dots \ . \end{aligned}$$

Marshall and Shaked (1983a) proved:

Theorem 7.8. If in Model 7.7

(i) $B_1, B_2, \dots$ are NBU;

(ii) $X_1, X_2, \dots$ are independent;

(iii) $F_i(u)$ is decreasing in i for all $u \geq 0$;

(iv) The X_i's are independent of the B_i's;

(v) The deterministic decrease is as described in (7.1);

then Z is an NBU process.

Theorem 7.8 is a special case of a more general result of Marshall and Shaked (1983a). Further extensions were obtained by Shanthikumar (1983). In particular Shanthikumar allowed B_i and X_i to be dependent, $i = 1,2,\dots$.

In some applications one may encounter the following shock model which is, in some sense, a "dual" of Model 7.7:

Model 7.9. (i) Same as Model 7.2(i).

(ii) At time $t = 0$ the damage level of the device is 0. Upon occurrence of the ith shock the device "enjoys" a nonpositive random "damage" with distribution F_i (addition of a nonpositive damage means reduction of the damage level, but the damage level never becomes negative). "Damages" are independent of the renewal process in (i).

(iii) Same as Model 3.1(iii).

(iv) Between shocks the damage level increases in some deterministic fashion.

Here let $B_1, B_2, \ldots$ denote the intershock times of (i), $X_1, X_2, \ldots$ denote the nonpositive "damages" of (ii) and $\{Z(t), t \geq 0\}$ denote the associated damage process.

Usually the deterministic increase of Z between shocks is of the form "the rate of increase at time t depends only on the level of damage at t provided t is not an epoch in the renewal process in (i)". Thus, for some strictly increasing function $h:[0,\infty) \to [0,\infty)$ {here $R_0 \equiv 0$, $R_i = B_1 + \ldots + B_i$}

$$\begin{aligned} Z(0) &= 0 \\ Z(t) &= h(t-R_i+h^{-1}(Z(R_i))) \quad \text{if } R_i < t < R_{i+1},\ i = 0,1,2,\ldots \\ &= [Z(R_i-) + X_i]^+ \quad \text{if } t = R_i,\ i = 1,2,\ldots\ . \end{aligned} \tag{7.2}$$

Marshall and Shaked (1983a) proved:

<u>Theorem 7.10.</u> If in Model 7.9

(i) $B_1, B_2, \ldots$ are exponential;

(ii) $X_1, X_2, \ldots$ are independent;

(iii) $F_i(u)$ is decreasing in i for all $u \leq 0$;

(iv) The deterministic increase is as described in (7.2);

then Z is an NBU process.

Shanthikumar (1983) has extended Theorem 7.10 in various

ways. In particular, he allowed the B_i's in Theorem 7.10(i) to have a decreasing failure rate (DFR) distribution.

Ross (1979) has studied a stronger definition of an NBU process. So far his definition has not been used often. However, most of the processes, mentioned in this paper so far, are also NBU according to the definition of Ross (1979).

8. NBUE AND HNBUE PROCESSES

Combination of Lemma 4.1a of Esary, Marshall and Proschan (1973) with Theorem 2.3 of Block and Savits (1978) gives (using 2.1):

Theorem 8.1. If in Model 7.2

(i) $B_1, B_2, \ldots$ are NBUE;

(ii) $X_1, X_2, \ldots$ are independent;

(iii) $F_i(u)$ is decreasing in i for all $u \geq 0$;

then the associated damage process Z is NBUE.

Actually, from Theorem 2.3 of Block and Savits (1978) it follows that if the B_i's are independent (but not necessarily identically distributed) NBUE and $E[B_i]$ is decreasing in i, then the associated damage process is NBUE.

Sumita and Shanthikumar (1983) relaxed the assumption of independence of B_i and X_i. They proved:

Theorem 8.2. If in Model 7.4

(i) $B_1, B_2, \ldots$ are NBUE;

(ii) $(B_1, X_1), (B_2, X_2), \ldots$ are independent and identically distributed;

(iii) X_1 is right tail increasing in B_1 (i.e., $P\{X_1 > x \mid B_1 > b\}$ is increasing in b for all $x \geq 0$);

then the associated damage process is NBUE.

A related result which (together with Lemma 4.1a of Esary,

Marshall and Proschan (1973) and (2.1) of this paper) indicates situations which give rise to NBUE processes is Theorem 2.11 of A-Hameed and Proschan (1975).

Combination of Lemma 4.1a of Esary, Marshall and Proschan (1973) with Corollary 5.6 of Klefsjo (1981b) gives (using (2.1)):

<u>Theorem 8.3.</u> If in Model 7.2

(i) $B_1, B_2, \ldots$ are HNBUE;

(ii) $X_1, X_2 \ldots$ are independent;

(iii) $\bar{F}_i(u)$ is decreasing in i for all $u \geq 0$;

then the associated damage process Z is HNBUE.

Relaxing the assumption of independence of B_i and X_i in Theorem 8.3, but putting more stringent assumptions on the X_i's, Sumita and Shanthikumar (1983) proved:

<u>Theorem 8.4.</u> If in Model 7.4

(i) $B_1, B_2, \ldots$ are HNBUE;

(ii) $(B_1, X_1), (B_2, X_2), \ldots$ are independent and identically distributed;

(iii) $X_1, X_2, \ldots$ are exponential;

(iv) X_1 is right tail increasing in B_1;

then the associated damage process is HNBUE.

9. <u>MULTIVARIATE IFRA PROCESSES</u>

Let $\{\underline{Z}(t) = (Z_1(t), \ldots, Z_n(t)),\ t \geq 0\}$ be a nonnegative right-continuous vector-valued random process. The process $\underline{Z}$ is called a <u>multivariate IFRA process</u> if for every closed upper set $U \subset R^n$, the first passage time $T_U \equiv \inf\{t : \underline{Z}(t) \in U\}$ is IFRA (a set $U \subset R^n$ is called an upper set of its indicator function is increasing). Clearly, every component $\{Z_i(t),\ t \geq 0\}$ of a multivariate IFRA process is a univariate IFRA process.

Block and Savits (1981) introduced the above definition of

vector-valued IFRA processes. Among other things they proved or essentially proved the following results:

Theorem 9.1. If $\{\underline{Z}(t),\ t \geq 0\}$ is an increasing IFRA process and $\Psi_1,\ldots,\Psi_k$ are nonnegative increasing right-continuous functions then $\{\Psi_1(\underline{Z}\ (t)),\ldots,\Psi_k(\underline{Z}\ (t))),\ t \geq 0\}$ is an IFRA process.

Theorem 9.2. If $\{(Z_1(t),\ldots Z_n(t)),\ t \geq 0\}$ and $\{(Y_1(t),\ldots,Y_m(t)),\ t \geq 0\}$ are increasing independent IFRA processes then $\{(Z_1(t),\ldots Z_n(t),\ Y_1(t),\ldots,\ Y_m(t)),\ t \geq 0\}$ is an IFRA process.

Note that Theorem 6.4 follows at once from Theorems 9.1 and 9.2.

Theorem 9.3. The increasing nonnegative process $\{\underline{Z}\ (t),\ t \geq 0\}$ is an IFRA process if and only if, for every collection $U_1,\ldots,U_m$ of closed upper sets the joint distribution of $T_{U_1},\ldots,T_{U_m}$ satisfies condition B of Esary and Marshall (1979), that is, $\tau(T_{U_1},\ \ldots\ ,\ T_{U_m})$ is IFRA for every coherent life function τ (for the definition of coherent life functions see Barlow and Proschan (1975)).

Theorem 9.3 led Block and Savits (1981) to the following definition:

Definition 9.4. Let $\{\underline{Z}\ (t), t \geq 0\}$ be a nonnegative right-continuous increasing process. The process $\underline{Z}$ is called a strongly IFRA process if, for every collection $U_1,\ldots,U_m$ of closed sets the joint distribution of $T_{U_1},\ldots,T_{U_m}$ is multivariate IFRA (MIFRA) in the sense of Block and Savits (1980b), that is, if $g(T_{U_1},\ldots,T_{U_m})$ is IFRA for all nonnegative increasing subhomogeneous functions g (for the definition of, and a discussion on, subhomogeneous functions see Marshall and Shaked (1982)).

Every strongly IFRA process is also an IFRA process because

every coherent life function is subhomogeneous.

Marshall and Shaked (1979) have introduced the following multivariate cumulative damage shock model:

Model 9.5.

(i) n devices are subjected to shocks occurring randomly in time as events in a Poisson process.

(ii) At time $t = 0$ the damage level of each device is 0. Upon occurrence of the ith shock the devices suffer nonnegative random damages with joint distribution F_i. The damages are independent of the Poisson process in (i).

(iii) Each random damage, at the time of its occurrence, adds to the current damage level of the appropriate device.

(iv) Between shocks the damage level of each device stays constant.

Let $\{N(t), t \geq 0\}$ be the Poisson process in Model 9.5(i) and let $X_j^{(i)}$ be the damage suffered by the jth device at the time of the ith shock, so that $\underline{X}^{(i)} = (X_1^{(i)},\ldots,X_n^{(i)})$ is distributed according to F_i, $i = 1,2,\ldots$. Let $S_0^{(j)} \equiv 0$, $S_i^{(j)} = X_1^{(j)} + \ldots + X_i^{(j)}$, $i = 1,2, \ldots$; $j = 1,2,\ldots, n$. Finally, let $Z_j(t) = S_{N(t)}^{(j)}$. The process $\{(Z_1(t),\ldots,Z_n(t)), t \geq 0\}$ is called the damage process associated with Model 9.5.

Savits and Shaked (1981) showed the following:

Theorem 9.6. If in Model 9.5(ii)

(i) $\underline{X}^{(1)}, \underline{X}^{(2)},\ldots$ are independent;

(ii) $P\{\underline{X}^{(i)} \in U\}$ is increasing in i for every measurable upper set U;

then $\{(Z_1(t),\ldots,Z_n(t)), t \geq 0\}$ is an IFRA process.

Theorem 9.7. If in Model 9.5(ii)

(i) $\underline{X}^{(1)}, \underline{X}^{(2)}, \ldots$ are independent;

(ii) $F_1 = F_2 = \ldots$;

then $\{(Z_1(t),\ldots,Z_n(t)),\ t \geq 0\}$ is a strongly IFRA process.

10. MULTIVARIATE NBU PROCESSES

Let $\{\underline{Z}(t) = (Z_1(t),\ldots,Z_n(t)), t \geq 0\}$ be a nonnegative right-continuous vector-valued random process. The process $\underline{Z}$ is called a multivariate NBU process if for every closed upper set $U \subset R^n$, $T_U = \inf\{t:\underline{Z}(t) \in U\}$ is NBU.

El-Neweihi and Proschan (1982) and Marshall and Shaked (1983b) [see also El-Neweihi (1983)], which introduced this definition, obtained, among other things, the following analogs of Theorems 9.1-9.3 (actually, Marshall and Shaked (1983b) discuss processes in separable Banach spaces and not necessarily in R^n):

Theorem 10.1. If $\{\underline{Z}(t),\ t \geq 0\}$ is a (not necessarily monotone) NBU process and $\psi_1,\ldots,\psi_k$ are nonnegative increasing functions then $\{(\psi_1(\underline{Z}(t)),\ldots,\psi_k(\underline{Z}(t))), t \geq 0\}$ is an NBU process provided it is right-continuous.

Theorem 10.2. If $\{Z_1(t),\ldots,Z_n(t)),\ t \geq 0\}$ and $\{(Y_1(t),\ldots,Y_m(t)),\ t \geq 0$ are increasing independent NBU processes then $\{(Z_1(t),\ldots,Z_n(t),Y_1(t),\ldots,Y_m(t)),\ t \geq 0\}$ is an NBU process.

Note that Theorem 7.6 follows at once from Theorems 10.1 and 10.2.

Theorem 10.3. The increasing nonnegative process $\{\underline{Z}(t),\ t \geq 0\}$ is an NBU process if and only if, for every collection $U_1,\ldots,U_m$ of closed upper sets the joint distribution of $T_{U_1},\ldots,T_{U_m}$ satisfies the NBU analog of condition B of Esary and Marshall (1979).

In Section 9 we described a cumulative damage shock model

which gives rise to multivariate IFRA processes (see Theorem 9.6). The damage process associated with Model 9.5 is monotone (increasing). Some useful nonmonotone NBU processes have been obtained by Marshall and Shaked (1983b) from more general and realistic shock models. One of their models is the following:

Model 10.4 (i) n devices are subjectd to shocks occurring randomly in time as events in a renewal process.

(ii) Same as Model 9.5(ii).

(iii) Same as Model 9.5(iii).

(iv) Between shocks the damage levels decrease in some deterministic fashion (although damage levels never become negative).

For Model 10.4 Marshall and Shaked (1983b) obtained a multivariate analog of Theorem 7.8. Let $B_1, B_2, \ldots$ be the random intershock times in Model 10.4(i). Let $\underline{X}^{(i)} = (X_1^{(i)}, \ldots, X_n^{(i)})$ be the vector of random damages suffered by the devices at the time of the ith shock. Denote the associated damage process by $\{\underline{Z}(t) = (Z_1(t), \ldots, Z_n(t)),\ t \geqslant 0\}$. Assume that the deterministic decrease, between shocks, of the damage level of the jth device is determined by a strictly decreasing function g_j as in (7.1), so that (denoting $R_0 \equiv 0$, $R_i = B_1 + \ldots + B_i$)

$$
\begin{aligned}
Z_j(t) &= 0 \quad \text{if} \quad 0 \leqslant t < R_1 \\
&= [g_j(t-R_i + g_j^{-1}(Z_j(R_i)))]^+ \\
&\qquad \text{if} \quad R_i < t < R_{i+1},\ i = 1,2,\ldots, \qquad (10.1) \\
&= [Z_j(R_i-) + X_j^{(i)}]^+ \\
&\qquad \text{if} \quad t = R_i,\ i = 1,2,\ldots\ .
\end{aligned}
$$

Marshall and Shaked (1983b) essentially proved the following:

Theorem 10.5. If in Model 10.4

(i) $B_1, B_2, \ldots$ are NBU;

(ii) $\underline{X}^{(1)}, \underline{X}^{(2)}, \ldots$ are independent;

(iii) $P\{\underline{X}^{(i)} \in U\}$ is increasing in i for every measurable upper set U;

(iv) the deterministic decrease is as described in (10.1);

then $\underline{Z}$ is an NBU process.

Marshall and Shaked (1983b) also obtained a Banach space analog of Theorem 7.10.

We end this section by mentioning that there exists, in the literature, an earlier occurrence of a multivariate NBU process arising from a cumulative damage shock model. Ross, Shahshahani and Weiss (1980), Proposition 4, have described a discrete cumulative damage shock model (all damages are either 0 or 1 and shocks occur at every time point which is a positive integer) and proved that the associated damage process is multivariate NBU. It should be noted that this result also follows from Theorem 10.5 by taking the B_i's degenerate at 1, the $\underline{X}^{(i)}$'s independent identically distributed random vectors which can take on values only on the vertices of the n-dimensional unit cube, and the damage levels staying constant between shocks.

References

Abdel-Hameed, M. (1975). A gamma wear process. IEEE Transactions on Reliability, R-24, 152-153.

Abdel-Hameed, M. (1983). Pure jump damage processes. Technical report, Department of Mathematics, University of North Carolina at Charlotte.

A-Hameed, M. S. and Proschan, F. (1973). Nonstationary shock models. Stochastic Processes and Their Applications, 1, 383-404.

A-Hameed, M. S. and Proschan, F. (1975). Shock models with underlying birth process. J. Appl. Prob., 12, 18-28.

Assaf, D., Shaked, M. and Shanthikumar, J. G. (1983). First passage times with PF_r densities. Technical report, Department of Mathematics, University of Arizona.

Barlow, R. E. and Proschan, F. (1975). Statistical Theory of Reliability and Life Testing: Probability Models. Holt, Rinehart and Winston, New York.

Block, H. W. and Savits, T. H. (1978). Shock models with NBUE survival. J. Appl. Prob., 15, 621-628.

Block, H. W. and Savits, T. H. (1980a). Laplace transforms for classes of life distributions. The Annals of Probability, 8, 465-474.

Block, H. W. and Savits, T. H. (1980b). Multivariate increasing failure rate average distributions. The Annals of Probability, 8, 793-801.

Block, H. W. and Savits, T. H. (1981). Multidimensional IFRA processes. The Annals of Probability, 9, 162-166.

El-Neweihi, E. (1983). Classes of multivatiate new better than used processes. The Institute of Mathematical Statistics Bulletin, 12, 85.

El-Neweihi, E. and Proschan, F. (1982). Degradable systems: A Survey of multistate system theory. Technical report, Department of Statistics, Florida State University.

El-Neweihi, E., Proschan, F. and Sethuraman, J. (1978). Multistate coherent systems. J. Appl. Prob., 15, 675-688.

Esary, J. D. and Marshall, A. W. (1979). Multivariate distributions with increasing hazard rate average. The Annals of Probability, 7, 359-370.

Esary, J. D., Marshall, A. W. and Proschan, F. (1973). Shock models and wear processes. The Annals of Probability, 1, 627-643.

Ghosh, M. and Ebrahimi, N. (1982). Shock models leading to increasing failure rate and decreasing mean residual life survival. J. Appl. Prob., 19, 158-166.

Gottlieb, G. (1980). Failure distributions of shock models. J. Appl. Prob., 17, 745-752.

Klefsjo, B. (1981a). Survival under the pure birth shock model. J. Appl. Prob., 18, 554-560.

Klefsjo, B. (1981b). HNBUE survival under some shock models. Scand. J. Statist., 8, 39-47.

Marshall, A. W. and Shaked, M. (1979). Multivariate shock models for distributions with increasing hazard rate average. The Annals of Probability, 7, 343-358.

Marshall, A. W. and Shaked, M. (1982). A class of multivariate new better than used distributions. The Annals of Probability 10, 259-264.

Marshall, A. W. and Shaked, M. (1983a). New better than used processes. Adv. Appl. Prob., 15, 601-615.

Marshall, A. W. and Shaked, M. (1983b). NBU processes with general state space. Technical report, Department of Mathematics, University of Arizona.

Ross, S. M. (1979). Multivalued state components systems. The Annals of Probability, 7, 379-383.

Ross, S. M. (1981). Generalized Poisson shock models. The Annals of Probability, 9, 896-898.

Ross, S. M., Shahshahani, M. and Weiss, G. (1980). On the number of component failures in systems whose component lives are exchangable. Mathematics of Operations Research, 5, 358-365.

Savits, T. H. and Shaked, M. (1981). Shock models and the MIFRA property. Stochastic Processes and Their Applications, 11, 273-283.

Shanthikumar, J. G. (1983). Processes with new better than used first passage times. To appear in Advances in Applied Probability.

Sumita, V. and Shanthikumar, J. G. (1983). General cumulative shock models. Technical report, The Graduate School of Management, The University of Rochester.

STOCHASTIC PROPERTIES OF A SEQUENCE OF INTERFAILURE TIMES UNDER MINIMAL REPAIR AND UNDER REVIVAL

Harold S. Balaban

ARINC Research Corporation
Annapolis, Maryland

Nozer D. Singpurwalla[1]

Department of Operations Research
The George Washington University
Washington, D.C.

Most of the literature on the topic of interfailure times following repair is based on the assumption of maximal repair; *that is, a failed item is restored to a condition equal to that of a new item. For some applications, the validity of this assumption has been challenged. In this paper we study the situation wherein the failed item is assumed to have been restored to a condition equal to its condition just prior to failure. This is known as* minimal repair. *We contrast the two repair policies, cite some implications of minimal repair, and obtain some preservation properties. Finally, we draw attention to the fact that under certain conditions, minimal repair actions generate a class of survival functions which cannot be described by any of the well known properties of ageing studied in reliability theory.*

[1]*Research supported by the Office of Naval Research, Contract N00014-77-C-0263, and by the Naval Air Systems Command, Contract N00019-82-C-0458, with the George Washington University.*

ISBN 0-12-041420-1

1. INTRODUCTION

In this paper we attempt to describe the stochastic behavior of a sequence of random variables which arise in a study of repairable systems. Much of the published literature on this topic is based on the assumption that the repair of a failed item restores it to the status of a new item, so that the available machinery of renewal theory can be applied. Such perfect repair actions will be termed "maximal repair." In a series of papers, mostly published in the engineering literature, Ascher, 1968,1979, and Ascher and Feingold, 1978,1979, have questioned the reasonableness of maximal repair in many applications, and have thus questioned the value of the ensuing results. Their assertion is that in practice repair is imperfect, in the sense that a failed item is often restored to a condition which is the same as its condition *just prior* to failure, or at best, to a condition which is slightly better. Thus the need for a more realistic description of the stochastic behavior of interfailure times following repair is germane, and this paper is a preliminary effort towards that goal. Here, we shall focus attention on repair actions which restore a failed item to a condition equal to its condition just prior to failure. Barlow and Hunter, 1960, term such imperfect repair actions "minimal repair actions"; Ascher and Feingold, 1969, use the more colloquial expression "bad as old" to refer to the state of the item after repair. It is appropriate to mention here that in Lewis, 1964, and in Brown and Proschan, 1980,certain aspects of imperfect repair have been considered; however, their models, approaches, and the nature of their results are quite different from ours.

Examples of minimal repair actions are:

(1) A TV set has stopped functioning because of the failure of an integrated circuit (I-C) panel. The set functions as soon as the failed panel is replaced. If the other components are left

alone, the set is not like a new one; minimal repair has been performed.

(2) A tire which has several miles on it is punctured by a nail and goes flat; the vehicle using the tire is considered to have failed. A repair of the puncture restores the vehicle to an operational status. If we assume that the puncture patch has not strengthened the tire by a significant amount, then a minimal repair has been performed on the vehicle.

(3) A coronary occlusion may cause heart failure. Cardiopulmonary resuscitation (CPR) may revive the patient. Assuming revival without damage to vital organs, we may view CPR as minimal repair.

1.1 Notation and Preliminaries

Let Ω be a measure space endowed with a probability measure P defined on the class L of Borel sets in Ω. Let I denote the set of positive integers, and for every $n \in I$, we shall define a finite, nonnegative, and real valued function $X(n;\omega)$ which is a P-measurable function of $\omega \in \Omega$. In the interest of brevity, we write $X(n)$ instead of $X(n;\omega)$, and note that $\{X(n)\}$ is a stochastic process whose domains are the sets I and Ω.

For our development, we shall require that $X(n) \leqslant X(n+1)$, $n = 1,2,\ldots$; the motivation for this requirement will be clarified later.

For every fixed n, $X(n)$ is a random variable whose distribution function we shall denote by $F_{X(n)}(x)$. The survival function $P\{X(n) \geqslant x\}$ is denoted by $\bar{F}_{X(n)}(x) = 1 - F_{X(n)}(x)$.

For an arbitrary finite set of n-values, say $n = 1,2,\ldots,m$, the corresponding random variables $X(1),X(2),\ldots,X(m)$ will have a joint m-dimensional distribution, with distribution function

$$F_{X(1),\ldots,X(m)}(x_1,\ldots,x_m) = P\{X(1) \leqslant x_1,\ldots,X(m) \leqslant x_m\}.$$

The family of all these joint probability distributions for

$m = 1,2,\ldots,$ and all possible values of $(x_1,\ldots,x_m)$, constitutes the family of finite dimensional distributions associated with the $\{X(n)\}$ process. Since I is discrete, the family of finite dimensional distributions uniquely determines the probability that the point $X = (X(1),X(2),\ldots)$ belongs to any Borel set of $\mathbb{R}^{\infty}$.

We shall designate the life length of a new item by $X(1)$, and whenever there is no cause for confusion, write $F(x)$ for $F_{X(1)}(x)$. In what follows we present certain useful notions which are standard in reliability theory.

The *failure rate* $r(x) \stackrel{\text{def}}{=} \lim_{t\to 0} (F(x+t) - F(x))/t(1-F(x))$ is assumed to exist, and the *cumulative failure rate* $R(x) = \int_0^x r(u)du$ is related to the survival function $\bar{F}(x)$, by $\bar{F}(x) = \exp[-R(x)]$.

Since items subjected to repair and other maintenance actions are those which age or experience wearout, a characterization of wear (see Barlow and Proschan, 1975, Ch. 4) given in Definition 1 will be useful. The notation "$H(x) \uparrow(\downarrow) x$" denotes that the function $H(x)$ is nondecreasing (nonincreasing) in x.

Definition 1. A distribution F (or its survival function $\bar{F}$) with $F(0) = 0$ is

(a) IFR(DFR) if $\bar{F}(x+t)/\bar{F}(x) \downarrow(\uparrow) x$, for $x \geqslant 0$ and each $t \geqslant 0$;
(b) IFRA(DFRA) if $-x^{-1}\log\bar{F}(x) \uparrow(\downarrow) x$, for $x > 0$;
(c) DMRL(IMRL) if $\int_x^{\infty} \bar{F}(t|x)dt \downarrow(\uparrow) x$, for $x \geqslant 0$;
(d) NBU(NWU) if $\bar{F}(x+y) \leqslant(\geqslant) \bar{F}(x)\bar{F}(y)$, for $x,y \geqslant 0$;
(e) NBUE(NWUE) if $\int_0^{\infty} \bar{F}(t+x|x)dt \leqslant(\geqslant) \int_0^{\infty} \bar{F}(t)dt$, for $x \geqslant 0$;

$\bar{F}(t|x)$ denotes the conditional reliability of a unit of age x.

The notation "$A \Rightarrow B$" denotes that A implies B. The following chain of implications is well known (see, for example, Haines and Singpurwalla, 1974, p. 62):

$$\text{NBUE} \Leftarrow \text{DMRL} \Leftarrow \text{IFR} \Rightarrow \text{IFRA} \Rightarrow \text{NBU} \Rightarrow \text{NBUE}$$
$$\text{NWUE} \Leftarrow \text{IMRL} \Leftarrow \text{DFR} \Rightarrow \text{DFRA} \Rightarrow \text{NWU} \Rightarrow \text{NWUE}.$$

2. STOCHASTIC PROCESSES GENERATED BY MINIMAL AND MAXIMAL REPAIR

For convenience, we denote our collection of random variables $\{X(n);\ X(n+1) \geqslant X(n),\ n \in I\}$, where $X(n)$ is the time of the successive failure, by $\{X(n)\}$. Based upon considerations which describe minimal repair, we note

Property 1. The stochastic process $\{X(n)\}$, with $X(0) \overset{\text{def}}{=} 0$, generated by a minimal repair policy is such that for all $n \in I$,

$$\begin{aligned} &P\{X(n+1) \geqslant y \mid X(n) = x,\ X(n-1) = z, \ldots\} \\ &= P\{X(n+1) \geqslant y \mid X(n) = x\} = P\{X(1) \geqslant y \mid X(1) \geqslant x\} \end{aligned}$$

for all $z < x, \ldots,$ and all $0 \leqslant x < y < \infty$.

Property 1 states that for each $n \in I$, the distribution of life length following the nth failure, given that the nth failure occurred at x, is the same as the distribution of the first life length given that it was at least x. This characterizes a minimal repair action, for such an action restores the item to the operable state it was in just before failure. In effect, the failure and the associated maintenance action have no discernible effect on the ageing process.

It follows from the above (also see Blumenthal, Greenwood, and Herbach, 1976), that the counting process generated by a minimal repair policy is a nonhomogeneous Poisson process having an intensity rate equal to the failure rate of the life distribution F. Specifically, if $N(x,t]$ denotes the number of minimal repair actions that occur in $(x,t]$, $0 \leqslant x < t$, then for all $m \in I$,

$$P\{N(x,t] = m\} = e^{-(R(t)-R(x))}(R(t) - R(x))^m / m!.$$

In contrast to minimal repair we have maximal repair, wherein a failed item is either totally overhauled so that it becomes as good as new, or better still, replaced by a new unit. For this we have

Property 2. The stochastic process $\{X(n)\}$, with $X(0) \overset{\text{def}}{=} 0$, generated by a maximal repair policy is such that for all $n \in I$,

$$\begin{aligned} &P\{X(n+1) \geqslant y \mid X(n) = x,\ X(n-1) = z, \ldots\} \\ &= P\{X(n+1) \geqslant y \mid X(n) = x\} = P\{X(1) \geqslant y - x\} \end{aligned}$$

for $z < x, \ldots,$ and all $0 \leqslant x < y < \infty$.

Note that a maximal repair policy generates an ordinary renewal process. Also, both the minimal and the maximal repair policies exhibit a Markov property.

2.1 *A Characterization of the Exponential Distribution*

We use the notation of Section 1, and let $\bar{F}_{X(n+1)|X(n)}(y|x)$ denote $P\{X(n+1) \geqslant y \mid X(n) = x\}$. Then, under a minimal repair policy, $\{X(n)\}$ is such that

$$\bar{F}_{X(n+1)|X(n)}(y|x) = \begin{cases} 1 & , \text{ for } 0 \leqslant y \leqslant x \\ \dfrac{\bar{F}_{X(1)}(y)}{\bar{F}_{X(1)}(x)}, & \text{for } 0 \leqslant x < y < \infty \end{cases} \tag{2.1}$$

and $\bar{F}_{X(1)}(x) \neq 0$, and under a maximal repair policy $\{X(n)\}$ is such that

$$\bar{F}_{X(n+1)|X(n)}(y|x) = \bar{F}_{X(1)}(y-x), \text{ for } 0 \leqslant x \leqslant y < \infty. \tag{2.2}$$

The exponential distribution has played a unique role in reliability theory, by virtue of the fact that in Definition 1, it is the only distribution which is a member of every defining class and its dual. For example, the exponential distribution is both IFR and DFR. An analogous role is played by this distribution with respect to minimal and maximal repair policies. We state this more precisely in the following characterization theorem.

Theorem 1. The stochastic process $\{X(n)\}$ satisfies both (2.1) and (2.2) if and only if $\bar{F}_{X(1)}$ is an exponential survival function.

3. PRESERVATION PROPERTIES OF A MINIMAL REPAIR POLICY

For items subjected to minimal repair, interest often centers around the conditional interfailure time $Y(n+1)$ given $X(n)$, where $\{Y(n+1) = X(n+1) - X(n)\}$ and $X(n)$ is the time of last failure/restoration. Of particular interest are the ageing characteristics of $Y(1) = X(1)$, the time to failure of a fresh unit. Our goal here is to describe those characteristics of $X(1)$ that are preserved by $Y(n+1)|X(n)$ when a minimal repair policy is in effect. Equation (2.1) implies that $\{Y(n)\}$ is such that

$$\bar{F}_{Y(n+1)|X(n)}(y|x) = \bar{F}_{X(n+1)|X(n)}(y+x|x) = \frac{\bar{F}_{X(1)}(y+x)}{\bar{F}_{X(1)}(x)} .$$

In what follows, we adopt the convention that a property, say π, holds for the survival function $\bar{F}_{Y(n+1)|X(n)}(y|x)$, if and only if it holds for all values of x. We introduce

Definition 2. A class of distributions, say S, is said to be *closed under minimal repair*, if $\{X(n)\}$ satisfies (2.1), and

$$\bar{F}_{X(1)} \in S \Rightarrow \bar{F}_{Y(n+1)|X(n)} \in S, \quad \forall\, n \in I.$$

When the above implication fails to hold, we shall say that S is *not closed under minimal repair*.

Theorem 2, given next, is a key result of this paper. Figure 1 illustrates the implications of the theorem for the IFR class and its associated chain of implications. If an implication is not shown, then there exists an example to show that it is not true.

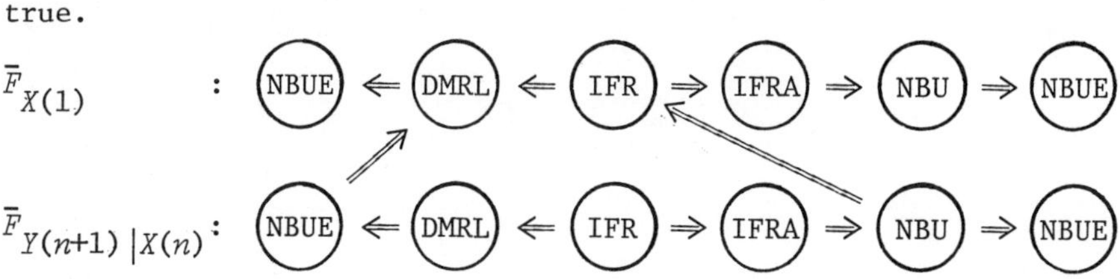

FIGURE 1. Chain indicating which of the ageing characteristics of $\bar{F}_{X(1)}$ are preserved by $\bar{F}_{Y(n+1)|X(n)}$.

Theorem 2.

(a) The IFR and DMRL classes are closed under minimal repair;

(b) The IFRA, NBU, and NBUE classes are not closed under minimal repair.

Analogous results hold for the DFR class and its associated chain of implications.

Proof: The proof consists of the series of lemmas given below.

Lemma 1. Let $\{X(n)\}$ satisfy (2.1). Then for all $n \varepsilon I$,

(a) $\bar{F}_{X(1)}$ IFR(DFR) $\Rightarrow \bar{F}_{Y(n+1)|X(n)}$ IFR(DFR),
(b) $\bar{F}_{Y(n+1)|X(n)}$ NBU(NWU) $\Rightarrow \bar{F}_{X(1)}$ IFR(DFR),
(c) $\bar{F}_{Y(n+1)|X(n)}$ NBUE(NWUE) $\Rightarrow \bar{F}_{X(1)}$ DMRL(IMRL).

Proof: $X(n)$ satisfying (2.1) $\Rightarrow$ for any $\delta > 0$, and $y \geqslant 0$,

$$\bar{F}_{Y(n+1)\ X(n)}(y+\delta|x) = \frac{\bar{F}_{X(1)}(y+x+\delta)}{\bar{F}_{X(1)}(x)}.$$

Thus

$$\frac{\bar{F}_{Y(n+1)|X(n)}(y+\delta|x)}{\bar{F}_{Y(n+1)|X(n)}(y|x)} = \frac{\bar{F}_{X(1)}(y+x+\delta)}{\bar{F}_{X(1)}(x)} \cdot \frac{\bar{F}_{X(1)}(x)}{\bar{F}_{X(1)}(y+x)}.$$

However,

$$\bar{F}_{X(1)}\,\text{IFR(DFR)} \Rightarrow \frac{\bar{F}_{X(1)}(y+x+\delta)}{\bar{F}_{X(1)}(y+x)} \downarrow(\uparrow)\ y,$$

and this establishes part (a) of the lemma. To prove part (b), let us assume that $\bar{F}_{X(1)}$ is not IFR. Then for some x, t, and δ,

$$\frac{\bar{F}_{X(1)}(x+\delta+t)}{\bar{F}_{X(1)}(x+\delta)} > \frac{\bar{F}_{X(1)}(x+t)}{\bar{F}_{X(1)}(x)}.$$

Rearranging, dividing both sides of the above inequality by $\bar{F}_{X(1)}(x)$, and using the fact that $\{X(n)\}$ satisfies (2.1), we have

$$\bar{F}_{X(n+1)\,|\,X(n)}(x+\delta+t\,|\,x) > \bar{F}_{X(n+1)\,|\,X(n)}(x+t\,|\,x)\bar{F}_{X(n+1)\,|\,X(n)}(x+\delta\,|\,x).$$

Thus $\bar{F}_{Y(n+1)\,|\,X(n)}$ is not NBU, and the result follows.

To prove part (c), we note that if $\bar{F}_{Y(n+1)\,|\,X(n)}$ is NBUE(NWUE), then for $\tau \geqslant 0$,

$$\int_0^\infty \frac{\bar{F}_{Y(n+1)\,|\,X(n)}(\tau+y\,|\,x)}{\bar{F}_{Y(n+1)\,|\,X(n)}(y\,|\,x)}\,d\tau \leqslant(\geqslant) \int_0^\infty \bar{F}_{Y(n+1)\,|\,X(n)}(\tau\,|\,x)d\tau$$

So

$$\int_0^\infty \frac{\bar{F}_{X(n+1)\,|\,X(n)}(\tau+y+x\,|\,x)}{\bar{F}_{X(n+1)\,|\,X(n)}(y+x\,|\,x)}\,d\tau \leqslant(\geqslant) \int_0^\infty \bar{F}_{X(n+1)\,|\,X(n)}(\tau+x\,|\,x)d\tau,$$

which implies

$$\int_0^\infty \frac{\bar{F}_{X(1)}(\tau+y+x)}{\bar{F}_{X(1)}(y+x)}\,d\tau \leqslant(\geqslant) \int_0^\infty \frac{\bar{F}_{X(1)}(\tau+x)}{\bar{F}_{X(1)}(x)}\,d\tau$$

and, hence,

$$\int_{y+x}^\infty \bar{F}_{X(1)}(\tau\,|\,y+x)d\tau \leqslant(\geqslant) \int_x^\infty \bar{F}_{X(1)}(\tau\,|\,x)\,d\tau,$$

which establishes that $X(1)$ is DMRL(IMRL).

To show closure under minimal repair of the DMRL and the IMRL classes of life distributions, we shall state and prove

Lemma 2. Let $\{X(n)\}$ satisfy (2.1). Then, for all $n \in I$,

$\bar{F}_{X(1)}$DMRL(IMRL) $\Rightarrow \bar{F}_{Y(n+1)\,|\,X(n)}$DMRL(IMRL).

Proof: $\bar{F}_{X(1)}$DMRL(IMRL) $\Rightarrow$

$$\int_0^\infty \frac{\bar{F}_{X(1)}(t+y+x)}{\bar{F}_{X(1)}(y+x)}\,dt \leqslant(\geqslant) \int_0^\infty \frac{\bar{F}_{X(1)}(t+y+\delta+x)}{\bar{F}_{X(1)}(y+\delta+x)}\,dt$$

for all $t \geqslant 0$ and $\delta \geqslant 0$. But then

$$\int_0^\infty \frac{\bar{F}_{X(1)}(t+y+x)}{\bar{F}_{X(1)}(x)} \bigg/ \left[\frac{\bar{F}_{X(1)}(y+x)}{\bar{F}_{X(1)}(x)}\right] dt$$

$$\leqslant(\geqslant) \int_0^\infty \frac{\bar{F}_{X(1)}(t+y+\delta+x)}{\bar{F}_{X(1)}(x)} \bigg/ \left[\frac{\bar{F}_{X(1)}(y+\delta+x)}{\bar{F}_{X(1)}(x)}\right] dt$$

and hence,

$$\int_0^\infty \frac{\bar{F}_{Y(n+1)|X(n)}(t+y|x)}{\bar{F}_{Y(n+1)|X(n)}(y|x)}\, dt \leqslant(\geqslant) \int_0^\infty \frac{\bar{F}_{Y(n+1)|X(n)}(t+\delta+y|x)}{\bar{F}_{Y(n+1)|X(n)}(y+\delta|x)}.$$

But this implies that $\bar{F}_{Y(n+1)|X(n)}(\cdot|x)$ is DMRL(IMRL) for each x.

Using Figure 1 and its DFR class counterpart, together with the well known facts that IFRA $\not\Rightarrow$ IFR and DFRA $\not\Rightarrow$ DFR, we can easily establish that the IFRA and DFRA classes are not closed under minimal repair. However, we have the following much stronger result.

Lemma 3. Let $\{X(n)\}$ satisfy (2.1). Then for all $n \in I$, $\bar{F}_{X(1)}\text{IFRA(DFRA)} \not\Rightarrow \bar{F}_{Y(n+1)|X(n)}\text{NBUE(NWUE)}$.

The details of the proof of this lemma are given in Balaban, 1978. For the IFRA class it rests on the following counterexample:

$$\bar{F}_{X(1)}(x) = \begin{cases} 1 & , 0 \leqslant x < 1 \\ e^{2(1-x)} & , 1 \leqslant x < 2 \\ e^{-x} & , 2 \leqslant x. \end{cases}$$

For the DFRA class the following counterexample is used:

$$\bar{F}_{X(1)}(x) = \begin{cases} e^{-x} & , 0 \leqslant x < 1 \\ e^{-\sqrt{x}} & , 1 \leqslant x < 4 \\ e^{-x/2} & , 4 \leqslant x. \end{cases}$$

This lemma has important practical consequences, since it establishes that IFRA(DFRA) life lengths under the action of minimal repair generate conditional distributions whose properties cannot be described by any of the well known notions of ageing studied in reliability theory. More about this will be said in Section 5.

Nonclosure of the NBU (NWU) and the NBUE (NWUE) classes under minimal repair is proved in Lemma 4, part (a) of which is a more general statement of lack of closure.

Lemma 4. Let $\{X(n)\}$ satisfy (2.1). Then, for all $n \in I$,

(a) $\bar{F}_{X(1)}\text{NBU(NWU)} \not\Rightarrow \bar{F}_{Y(n+1)|X(n)}\text{NBUE(NWUE)}$,

(b) $\bar{F}_{X(1)}\text{NBUE(NWUE)} \not\Rightarrow \bar{F}_{Y(n+1)|X(n)}\text{NBUE(NWUE)}$.

Proof: We prove part (a) by contradiction; the proof of part (b) is almost identical to the proof of part (a). Suppose that

$$\bar{F}_{X(1)}\text{NBU(NWU)} \Rightarrow \bar{F}_{Y(n+1)|X(n)}\text{NBUE(NWUE)}.$$

But $\bar{F}_{Y(n+1)|X(n)}\text{NBUE(NWUE)} \Rightarrow \bar{F}_{X(1)}\text{DMRL(IMRL)}$, by Lemma 1(c), and $\bar{F}_{X(1)}\text{DMRL(IMRL)} \Rightarrow \bar{F}_{Y(n+1)|X(n)}\text{DMRL(IMRL)}$ by Lemma 2. Thus we have

$$\bar{F}_{X(1)}\text{NBU(NWU)} \Rightarrow \bar{F}_{Y(n+1)|X(n)}\text{DMRL(IMRL)}.$$

To see that the above arguments lead us to a contradiction, we make use of Lemma 1(b) to establish that $\bar{F}_{X(1)}\text{IFRA(DFRA)} \Rightarrow \bar{F}_{X(1)}\text{NBU(NWU)} \Rightarrow \bar{F}_{Y(n+1)|X(n)}\text{IFR(DFR)} \Rightarrow \bar{F}_{Y(n+1)|X(n)}\text{IFRA(DFRA)}$, which according to Lemma 3 is false.

The proof of Theorem 2 is now complete.

Another counterexample which will immediately show nonclosure of the IFRA, NBU, and NBUE classes under minimal repair, is provided by the survival function of a parallel system of two independent components, each having an exponential survival function with scale parameter λ_i, $i = 1,2$, and $\lambda_1 \neq \lambda_2$. However, this counterexample cannot be used to show nonclosure of the DFRA, NWU, and NWUE classes, and thus the need for Lemmas 3 and 4.

4. A BOUND FOR THE CONDITIONAL SURVIVAL FUNCTION

The preservation properties of Theorem 2 enable us to take advantage of the several known bounds and inequalities of reliability theory (Barlow and Proschan, 1975, p. 109) for $\bar{F}_{Y(n+1)|X(n)}$, once we know its ageing characteristics. For example, bounds on $\bar{F}_{Y(n+1)|X(n)}$ can be obtained once we know that it is IFR (DFR). From Lemma 1, we also note that for all $n \in I$, $\bar{F}_{Y(n+1)|X(n)}$ IFR(DFR) $\Rightarrow \bar{F}_{Y(n+1)|X(n)}$ NBUE(NWUE) $\Rightarrow \bar{F}_{X(1)}$ IFR(DFR), so that $\bar{F}_{Y(n+1)|X(n)}$ is IFR (DFR) if and only if $\bar{F}_{X(1)}$ is IFR(DFR).

However, the fact that $\bar{F}_{X(1)}$ IFRA(DFRA) $\not\Rightarrow \bar{F}_{Y(n+1)|X(n)}$ NBUE (NWUE)(see Lemma 3) motivates us to develop some new bounds for $\bar{F}_{Y(n+1)|X(n)}$. This is particularly germane since IFRA lifetimes can arise quite naturally in practice. For example, if we consider the coronary occlusion situation of Section 1, and treat each occlusion as a shock to the heart muscle, then from the theory of shock models and wear processes (Esary, Marshall, Proschan, 1973), it follows under some very general conditions that the time to failure of the heart muscle is IFRA. If a revival of the patient by CPR can be regarded as a minimal repair action, then by Lemma 3, the time to the second failure of the heart, conditional on the time to the first failure, cannot be described by any of the notions of ageing given in Definition 1. This motivates us to describe the life length $Y(n+1)|X(n)$ under the assumption of minimal repair, and under the assumption that $X(1)$ is IFRA, a task presently under investigation. A useful upper(lower) bound on $\bar{F}_{Y(n+1)|X(n)}$ motivated by the IFRA (DFRA) property of $\bar{F}_{X(1)}$ is given in Theorem 3.

Theorem 3. Let $\{X(n)\}$ satisfy (2.1), and let $\bar{F}_{X(1)}$ be IFRA (DFRA). Then, for all $n \in I$, $0 < y < \infty$ and $x \geqslant 0$,

$$\bar{F}_{Y(n+1)|X(n)}(y|x) \leqslant(\geqslant) [\bar{F}_{X(1)}(y+x)]^{(y)/(y+x)}.$$

Proof: $\bar{F}_{X(1)}$ IFRA(DFRA) $\Rightarrow$

$$\bar{F}_{X(1)}(\alpha x) \leqslant(\geqslant) [\bar{F}_{X(1)}(x)]^{\alpha}, \text{ for } 0 < \alpha < 1.$$

Thus

$$\bar{F}_{Y(n+1)|X(n)}(y|x) = \frac{\bar{F}_{X(1)}(y+x)}{\bar{F}_{X(1)}(x)} = \frac{\bar{F}_{X(1)}(y+x)}{\bar{F}_{X(1)}(\alpha(y+x))} = \frac{\bar{F}_{X(1)}(y+x)}{\bar{F}_{X(1)}(\alpha(y+x))}$$

for $\alpha = (x/(y+x))$. By the defining property of IFRA(DFRA),

$$\bar{F}_{Y(n+1)|X(n)}(y|x) \leqslant(\geqslant) \frac{\bar{F}_{X(1)}(y+x)}{[\bar{F}_{X(1)}(y+x)]^{\alpha}} = [\bar{F}_{X(1)}(y+x)]^{1-\alpha},$$

or that

$$\bar{F}_{Y(n+1)|X(n)}(y|x) \leqslant(\geqslant) [\bar{F}_{X(1)}(y+x)]^{y/(y+x)}.$$

5. COMPARISON OF THE STOCHASTIC PROCESSES GENERATED BY MINIMAL AND MAXIMAL REPAIR POLICIES

The ageing properties of a new item provide us with a vehicle for comparing the relative desirability of a minimal repair policy versus a maximal repair policy. This in turn helps us choose a particular maintenance policy. Specifically, for an item which ages, a maintenance policy involving a complete overhaul or replacement is more desirable than one involving minimal repair, provided that cost and resource considerations are put aside. The converse is true for items which improve with age. More formally, we have:

Theorem 4. If $\bar{F}_{X(1)}$ is NBU(NWU), then for all $i \in I$,

$$P[Y(i+1) \geqslant t|X(i) = x \text{ and } \{X(n)\} \text{ satisfying (2.1)}]$$
$$\leqslant(\geqslant) P[Y(i+1) \geqslant t|X(i) = x \text{ and } \{X(n)\} \text{ satisfying (2.2)}].$$

Proof: Obvious, since the left-hand side of the above is $\bar{F}_{X(1)}(t+x)/\bar{F}_{X(1)}(x)$, and the right-hand side of the above is $\bar{F}_{X(1)}(t)$.

In order to develop some properties of joint life lengths occurring under minimal and maximal repair policies, we shall first present a basic definition pertaining to a general relationship among independent and certain types of nonindependent life lengths.

Definition 3 (Barlow and Proschan, 1975, p. 29). Random variables $X = (X_1,\ldots,X_n)$ are *associated* if

$$\mathrm{Cov}[\Gamma(X),\Delta(X)] \geqslant 0$$

for all pairs of binary increasing functions Γ and Δ.

Association implies a positive dependence among random variables; in the context of reliability theory, this is often realistic. Conditions for the association of interfailure/restoration times $Y(i)$, $i = 1,\ldots,n$ are given in Theorem 5 below.

Theorem 5. The sequence $\{Y(i)\}$, $i = 1,\ldots,n$ is associated if

(a) $\{X(n)\}$ satisfies (2.2)

(b) $\{X(n)\}$ satisfies (2.1) and $\bar{F}_{X(1)}$ is DFR.

Proof: Part (a) follows trivially from the fact that when $\{X(n)\}$ satisfies (2.2), the $Y(i)$'s are independent, and independent random variables are associated. To prove (b), we can show that under the conditions of the theorem, $Y(1),\ldots,Y(n)$ are *conditionally nondecreasing in sequence* (see Barlow and Proschan, 1975, p. 146), and are therefore associated.

ACKNOWLEDGMENTS

The authors are deeply grateful to Professors Frank Proschan, Bengt Klefsjo, Moshe Shaked, and a referee for going through earlier versions of this paper and making invaluable suggestions.

REFERENCES

ASCHER, H. (1968). Evaluation of repairable system reliability using the bad as old concept. *IEEE Trans. Reliability R-17*, 105-110.

ASCHER, H. (1979). Comments on "Models for reliability of repaired equipment." *IEEE Trans. Reliability R28*, 2, 119.

ASCHER, H. and FEINGOLD, H. (1969). "Bad as old" analysis of system failure. *Ann. Assurance Sci., 8th R&M Conference*. Gordon and Breech, New York, 49-62.

ASCHER, H. and FEINGOLD, H. (1978). Is there repair after failure? *Ann. Reliability and Maintainability Symp.*, IEEE Cat. No. 77CH13087R, 190-197.

ASCHER, H. and FEINGOLD, H. (1979). The aircraft air conditioner date revisited. *Proc. Ann. Reliability and Maintainability Symp.*, 153-159.

BALABAN, H. (1978). A stochastic characterization of failure processes under minimal repair. D.Sc. dissertation, George Washington University.

BARLOW, R. and HUNTER, M. (1960). Optimum preventive maintenance policies. *Operations Res. 8*, 90-100.

BARLOW, R. and PROSCHAN, R. (1975). *Statistical Theory of Reliability and Life Testing*. To Begin With, Silver Spring, MD.

BLUMENTHAL, S., GREENWOOD, J. A., and HERBACH, L. (1976). A comparison of the bad as old and superimposed renewal models. *Management Sci. 23*, 280-285.

BROWN, M. and PROSCHAN, F. (1980). Imperfect maintenance. IMS Lecture notes--Monograph Series, Vol. 2, *Survival Analysis* (J. Crowley and R. A. Johnson, eds.), 179-188.

ESARY, J. D., MARSHALL, A. W., and PROSCHAN, F. (1973). Shock models and wear processes. *Ann. Probability, 1*, 627-649.

HAINES, A. and SINGPURWALLA, N. D. (1974). Some contributions to the stochastic characterization of wear, in *Reliability and*

Biometry: Statistical Analysis of Life Length (F. Proschan and R. J. Serfling, eds.). Philadelphia: SIAM.

LEWIS, P. A. W. (1964). A branching Poisson process model for the analysis of computer failure patterns. *J. Roy. Statist. Soc. Ser. B*, 398-456.

Optimal Maintenance and Replacement Policies

SOME IMPERFECT MAINTENANCE MODELS

Robert A. Fontenot

Department of Mathematics
Whitman College
Walla Walla, Washington

Frank Proschan[1]

Department of Statistics
The Florida State University
Tallahassee, Florida

We develop optimum policies for several maintenance models based on the imperfect repair model of Brown and Proschan (J. Appl. Prob., 1983). In each of these we assume that planned replacement is perfectly executed. We assume further that unplanned maintenance is perfect with probability p and is imperfect (minimal repair) with probability 1 - p. We also assume that the cost of planned maintenance is higher than the cost of unplanned maintenance. In all our models we assume that after each unplanned repair there is an inspection requiring negligible time and yielding perfect information as to whether the repair is perfect or just minimal.

[1]*Research supported by the Air Force Office of Scientific Research under Grant AFOSR-82-K-0007.*

ISBN 0-12-041420-1

1. INTRODUCTION

During the past twenty-five years a large number of articles on maintenance policies have appeared. Most of these papers overlook two important factors in real-world maintenance operations: the possibility of errors on the part of the maintenance performer and limitations, physical or otherwise, which make complete overhaul of the unit needing repair impossible.

Recently, however, several authors have treated imperfect maintenance, that is, maintenance in which one or both of the factors just mentioned play a role. T. Nakagawa in [5] discusses several models in which the repaired unit never has effective age zero and several other models in which the maintenance performer accomplishes *planned* periodic maintenance perfectly (i.e., the repaired unit is as good as new) with probability p and performs only *minimal repair* (the unit is repaired so that it functions again, but has the same failure rate and the same effective age as at the time of failure) with probability $1 - p$. For the latter models, Nakagawa also assumes that *unplanned* maintenance, the repair of intermittent failures, is always perfect. Two other authors, M. Brown and F. Proschan, discuss general features of imperfect maintenance and inspection in [4] and develop properties of an imperfect repair model in [3]. For their imperfect repair model, the authors assume that *unplanned* repair is perfect with probability p and is minimal repair with probability $q = 1 - p$.

The main interest of the authors in [3] is in studying properties of the distribution F_p of the time between perfect repairs. For example, they show that the "failure rate" function of F_p is pr where r is the failure rate function of the life distribution F of the unit. They also show that if F belongs to certain life distribution classes based on various notions of aging, such as the classes of IFR and DFR distributions, then so does F_p. Finally, they prove stochastic inequalities concerned with F_p and related random variables.

Our concern in the present paper is different. We develop several maintenance optimization models based on the imperfect repair model of Brown and Proschan. In each of these we assume that *planned* replacement is always perfect. We assume further that unplanned maintenance is perfect with probability p and is imperfect (minimal repair) with probability $q = 1 - p$. We also assume that the cost of planned maintenance is higher than the cost of unplanned maintenance. Finally, in all of our models, we assume that after each unplanned repair there is an inspection, which takes negligible time and whose cost is included in the repair cost, which enables the operator of the unit to determine whether the repair was perfect or just minimal.

One possible interpretation of our models is the following. For unplanned maintenance the unit is repaired by a minimally trained or equipped repairman, whose fee is low; for planned maintenance, the unit is repaired by an extensively trained and well-equipped repairman, whose fee is rather high.

For information about minimal repair and other maintenance policies and for comparisons with our results in the imperfect maintenance setting, we refer the reader to Chapter 4 of [1].

We continue the introduction to this paper by giving notation that we shall use and assumptions that we shall make throughout this article. The life distribution of the unit under consideration will be denoted by F. The function $\overline{F}$ is $1 - F$. We assume throughout that F is absolutely continuous with density f and that $F(0) = 0$. The failure rate, $r(t) = f(t)\left(\overline{F}(t)\right)^{-1}$, is assumed to be continuous and increasing except in Theorem 5.2. The cumulative hazard function is

$$R(t) = \int_0^t r(u)du.$$

Note the relationship

$$\overline{F}(x) = e^{-R(x)}.$$

The distribution function of the time between successive perfect repairs, called F_p in [3], will be denoted by G. We shall use the fact, proved in [3], that $\overline{G} = \overline{F}^p$. The density of G is herein denoted by g. Other notations that we shall use are

$$r(\infty) = \lim_{t\to\infty} r(t),$$

$$\mu = \int_0^\infty \overline{F}(x)dx, \quad \gamma = \int_0^\infty \overline{F}^p(x)dx,$$

$$A(k) = \frac{1}{k!}\int_0^\infty e^{-R(x)}R^k(x)dx$$

for $k = 0, 1, 2, \ldots$. Note that $A(0) = \mu$. For convenience we shall assume that $p > 0$ throughout this paper. It will be seen that our results apply (in the limit as $p \to 0$) to the case $p = 0$ also; we merely wish to avoid having to write down separate formulas for the case $p = 0$. Finally, we assume that repairs and replacements or overhauls take negligible time.

In the remainder of this introductory section, we give a discussion of a basic model, which we call Model U, which naturally underlies the models considered in sections 2 through 5 of this paper.

Model U

A device has a life time distribution $F(t) = \int_0^t f(x)dx$ with f continuous. At the time of first failure of the device, a repair is performed. With probability p this repair is perfect. If it is perfect, the device is operated again with elapsed time to failure (that is, time between the first and second failures) given by the original distribution F. If the repair is imperfect (which happens with probability $q = 1 - p$), the device is operated with elapsed time to the next failure t_2 governed by the distribution with density $\dfrac{f(t_1 + t_2)}{\overline{F}(t_1)}$ where t_1 is time to first failure and $\overline{F}(t_1) = 1 - F(t_1)$. This process of repair and failure is repeated indefinitely. Notice that the elapsed time to the third failure will be governed by either $F(t_3)$, $f(t_2+t_3)/\overline{F}(t_2)$ or $f(t_1+t_2+t_3)/\overline{F}(t_1+t_2)$. We associate with this model the random variable X = total time to first perfect repair. As noted earlier, it is shown in [3] that the distribution function of X is $G(x) = 1 - \overline{F}^p(x)$.

Let $x > 0$. Associated with Model U, let N(x) denote the number of repairs in the interval $[0, x)$. The following lemma will be needed in the sequel.

1.1 Lemma. Let $x > 0$, $\delta > 0$ and k be a nonnegative integer.

(a) $$P(x < X < x+\delta \mid X > x) = \frac{\overline{G}(x) - \overline{G}(x+\delta)}{\overline{G}(x)}.$$

(b) $$P(N(x) = k,\ x < X < x+\delta) = \frac{\left(\overline{G}(x) - \overline{G}(x+\delta)\right)(qR(x))^k e^{-qR(x)}}{k!}.$$

(c) $$P\left(N(x) = k,\ X > x\right) = \frac{\left(qR(x)\right)^k e^{-R(x)}}{k!}.$$

Proof. Given that $X > x$, i.e., that no perfect repair has occurred in $[0, x)$, the (additional) time t to the next failure has density $h(t) = \dfrac{f(t+x)}{\overline{F}(x)}$. Using Lemma 2.1 in [3], we conclude that the distribution of the time (starting at x) to the first perfect repair is $\overline{H}^p(t)$ where H is the distribution function corresponding to the density h. An easy computation shows that $\overline{H}^p(t) = \dfrac{\overline{G}(x+t)}{\overline{G}(x)}$. Thus it follows that

$$P(x < X < x+\delta \mid X > x) = \overline{H}^p(x) - \overline{H}^p(x+\delta) = \frac{\overline{G}(x) - \overline{G}(x+\delta)}{\overline{G}(x)}.$$

(b) Consider the following subset of R^k:

$$E = \{(t_1, \ldots, t_k) : 0 \le t_1 < x,\ 0 \le t_i < x - (t_1 + \ldots + t_{i-1}),$$
$$1 \le i \le k\}.$$

Then it follows that

$$P\left(N(x) = k,\ x < X < x+\delta\right)$$

$$= q^k \int_E S(t_1, \ldots, t_k) \frac{\overline{F}(x)}{\overline{F}(t_1 + \ldots + t_k)} \frac{\overline{G}(x) - \overline{G}(x+\delta)}{\overline{G}(x)} dt_k \ldots dt_1$$

where

$$S(t_1, \ldots, t_k) = f(t_1) \prod_{i=1}^{k-1} \frac{f(t_1 + \ldots + t_{i+1})}{\overline{F}(t_1 + \ldots + t_i)} .$$

Rewriting, we obtain

$$\frac{q^k \overline{F}(x) [\overline{G}(x) - \overline{G}(x+\delta)]}{\overline{G}(x)} \int_E \prod_{i=1}^{k} r(t_1 + \ldots + t_i) dt_k \ldots dt_1 .$$

A straightforward calculation shows that this last integral has value $\frac{R^k(x)}{k!}$. Simplifying, we obtain the formula claimed in the statement of part (b).

(c) Use the result of (b), let $\delta \to \infty$, and then simplify the resulting expression.

2. MODEL I

Our first model, Model I, is a modified age replacement model. A device has life distribution F (with F having the properties specified in Section 1). At the beginning, a perfect repair is scheduled to take place at time T (a constant) at a cost of c_2. If the device fails at a time t_1 prior to T, it is repaired at a cost of $c_1 < c_2$. This repair is perfect with probability p and imperfect (minimal) with probability $q = 1 - p$. As we stated in Section 1, we assume that each unscheduled repair is followed by an inspection, taking negligible time and at no extra cost, to determine whether or not the repair was perfect. If the repair is perfect, the scheduled perfect maintenance session is rescheduled, with the same cost c_2, at a time T units of time in the future and the machine is put back into operation with elapsed time to the next failure governed by F. If the repair is imperfect, the device is put back into operation with elapsed time t_2 to the next failure governed by the density $\frac{f(t_1 + t_2)}{\overline{F}(t_1)}$. These steps are repeated at subsequent failures with appropriate modifications.

Clearly the process we have described renews itself at the time of the first perfect repair, scheduled or unscheduled. Also, there are obvious connections with Model U. For example, if Z is the elapsed time to the first perfect repair (scheduled or otherwise) associated with Model I and if X is the time to the first perfect repair associated with Model U, then $Z = \min\{X,T\}$.

The problem is to minimize L(T), the long-run expected cost per unit time. In Theorem 2.2, we give a formula for L(T), and in Theorem 2.3, we discuss the optimal value of T. Example 2.4 concerns the Weibull distribution. It is interesting to note that the formula for L(T) has the form of the long-run expected cost per unit time in a standard age replacement problem with life distribution G and costs $\frac{c_1}{p}$ and c_2. As a result we could deduce most or all of the results in Theorem 2.3 by translating known results into our notation.

2.1 Lemma. For the repair process associated with Model I, let N(t) denote the number of failures in [0, t), for t > 0, and let Z denote the time of the first perfect repair. Let $0 < x < T$, $\delta > 0$, and k be a nonnegative integer.

$$P\big(N(x) = k \mid Z = x\big) = \frac{\big(qR(x)\big)^k e^{-qR(x)}}{k!}.$$

Also, the conditional mean is qR(x).

Proof. Recall that g denotes the density of $G = 1 - \bar{F}^p$. Using Lemma 1.1 and the fact that $x < T$, we have that

$$\begin{aligned} P\big(N(x) = k \mid Z = x\big) &= P\big(N(x) = k \mid X = x\big) \\ &= \frac{\lim_{\delta\to 0} P\big(N(x) = k,\ x < X < x+\delta\big)\delta^{-1}}{\lim_{\delta\to 0} P(x < X < x+\delta)\delta^{-1}} \\ &= \frac{g(x)\big(qR(x)\big)^k e^{-qR(x)}(k!)^{-1}}{g(x)} = \frac{\big(qR(x)\big)^k e^{-qR(x)}}{k!}. \end{aligned}$$

Finally, we note that the mean of this Poisson random variable is qR(x).

2.2 Theorem.

$$L(T) = \frac{\frac{c_1}{p} G(T) + c_2\overline{G}(T)}{\int_0^T \overline{G}(x)dx} .$$

Proof. The times of perfect repair are renewal times in a renewal process whose interarrival distribution is that of $\min\{X,T\}$. From renewal theory, we have that

$$L(T) = \frac{C(T)}{D(T)} ,$$

where $C(T)$ is the expected cost per renewal cycle and $D(T)$ is the expected duration of a renewal cycle. It is clear that

$$D(T) = \int_0^T \overline{G}(x)dx .$$

From Lemma 2.1, it follows that

$$C(T) = c_1 \int_0^T \left(1 + qR(x)\right)g(x)dx + \left(c_1 qR(T) + c_2\right)\overline{G}(T) .$$

Noting that $R(x) = \frac{-\ell n\ \overline{G}(x)}{p}$ and using the substitution $u = \overline{G}(x)$, we obtain that

$$\int_0^T R(x)g(x)dx = -\overline{G}(T)R(T) + \frac{G(T)}{p} ,$$

from which it follows that

$$C(T) = \frac{c_1}{p} G(T) + c_2\overline{G}(T).$$

2.3 Theorem. In addition to the hypotheses given in Section 1, assume that r is differentiable.

Let

$$H(T) = (c_1 - pc_2)\left(r(T) \int_0^T \overline{G}(x)dx + \frac{1}{p}\overline{G}(T)\right) - \frac{c_1}{p} .$$

(a) $L'(T) = \overline{G}(T)\left(\int_0^T \overline{G}(x)dx\right)^{-2} H(T).$

(b) If $c_1 \leq pc_2$, the optimal replacement age is $T = \infty$.

(c) Assume that $c_1 > pc_2$. Then an optimal replacement age T^* exists (T^* may be infinite).

(d) The equation $H(T) = 0$ has at most one solution. If a solution exists, it must be T^*.

(e) Suppose that $c_1 > pc_2$ and that

$$r(\infty) > \frac{c_1}{p\gamma(c_1 - pc_2)} .$$

Then T^* is finite.

Proof. (a) Differentiate and use the identity $g = pr\overline{G}$.

(b) If $c_1 \leq pc_2$, $L'(T) < 0$ for $T > 0$.

(c) First note that H is increasing since $r(t)$ is increasing and an easy computation shows that

$$H'(T) = (c_1 - pc_2)r'(T) \int_0^T \overline{G}(x)dx.$$

Also note that $H(0) = -c_2 < 0$. Thus if the equation $H(T) = 0$ has a solution T^*, then L' is negative on $(0, T^*)$ and positive on (T^*, ∞). If $H(T) = 0$ has no solution, then L is decreasing and $T^* = \infty$.

(d) Look at $H'(T)$. Also note that if $H(T) = 0$ has a solution, then $c_1 > pc_2$.

(e) We must show that the equation $H(T) = 0$ has a solution. The inequality involving $r(\infty)$ gives us what we need since it implies that $\lim H(t) > 0$. (Recall that $H(0) < 0$.)

2.4 Example. Let F be an IFR Weibull distribution. Thus, for some $\lambda > 0$ and $\alpha \geq 1$,

$$\overline{F}(x) = e^{-\lambda x^\alpha}, \quad x \geq 0.$$

Assume that $c_1 > pc_2$. By Theorem 2.3, if $\alpha > 1$, a finite optimal replacement age T^* exists, since $r(\infty) = \infty$. T^* is the unique solution of the equation $H(T) = 0$. This equation reduces to the equation

$$\lambda\alpha T^{\alpha-1}\int_0^T e^{-p\lambda x^\alpha}dx + \frac{1}{p}e^{-p\lambda T^\alpha} = \frac{c_1}{p(c_1 - pc_2)},$$

which is easily solved by numerical methods (for specific λ and α).

If $\alpha = 1$, the equation above has no solution (except in the uninteresting case $c_2 = 0$). Thus, for the exponential distribution, the optimal replacement age is $T^* = \infty$.

3. MODEL II

In this section we discuss Model II, an imperfect repair version of a model developed in [6]. Model II is the same as Model I except in one respect: we replace (or perfectly repair) the unit on the next failure after $k-1$ successive imperfect repairs (instead of at times when T units of time have elapsed since the last perfect repair, as in Model I). The objective is to choose k so as to minimize L(k), the long-run expected cost per unit time.

The next result contains a formula for L(k). Recall the assumptions stated in Section 1.

3.1 Theorem.

$$L(k) = \frac{\frac{c_1}{p}(1-q^{k-1}) + c_2 q^{k-1}}{\sum_{j=0}^{k-1} q^j A(j)}.$$

Proof. From renewal theory, we have that

$$L(k) = \frac{C(k)}{D(k)},$$

where C(k) and D(k) are the expected cost per renewal cycle and the expected duration of a renewal cycle, respectively.

We compute C(k) first. A renewal cycle can end in one of two ways: (1) there are $k-1$ successive imperfect repairs and then another failure; or (2) there are $j-1$ successive imperfect repairs and then a perfect repair, for some integer j, $1 \leq j \leq k-1$. Considering the probabilities associated with these events, we obtain

$$C(k) = \left(c_1(k-1) + c_2\right)q^{k-1} + \sum_{j=1}^{k-1} c_1 jpq^{j-1} .$$

Simplifying, we get

$$C(k) = \frac{c_1}{p}(1 - q^{k-1}) + c_2 q^{k-1} .$$

Next we compute $D(k)$. Let $W(k)$ denote the length of time to the first renewal. Using Lemma 1.1, we have that

$$P\left(W(k) > t\right) = \sum_{j=0}^{k-1} P\left(X > t,\ N(t) = j\right) = \sum_{j=0}^{k-1} \frac{q^j [R(T)]^j}{j!} e^{-R(t)} .$$

Thus

$$D(k) = \int_0^\infty P\left(W(k) > t\right) dt = \sum_{j=0}^{k-1} q^j A(j).$$

To discuss the optimization problem associated with Model II, we need a lemma due to T. Nakagawa [6].

3.2 Lemma. The sequence $\left(A(k)\right)$ is decreasing and

$$\lim_{k\to\infty} A(k) = \frac{1}{r(\infty)} .$$

3.3 Theorem. Let $\beta = \dfrac{c_1 q}{c_1 - pc_2}$ and

$$Q(k) = \frac{pD(k)}{A(k)} + q^k$$

for $k = 1, 2, 3, \ldots$.

(a) The sequence $\left(Q(k)\right)$ is increasing.

(b) If $c_1 \le pc_2$, the sequence $\left(L(k)\right)$ is decreasing, and the optimal value of k is $k = \infty$.

(c) Suppose that $c_1 > pc_2$. Then for any positive integer k, $L(k+1) \ge L(k)$ if and only if $Q(k) \ge \beta$. Also, an optimal value of k, possibly infinity, exists.

(d) Suppose that $c_1 > pc_2$ and that $r(\infty) > \beta(p\gamma)^{-1}$. Then a finite optimal value k^* of k exists. Furthermore,

$$k^* = \min\{k\colon Q(k) \ge \beta\}.$$

Proof. (a) This follows from Lemma 3.2 and the identity $D(k+1) = D(k) + q^k A(k)$.

(b) It is easy to see that for any positive integer k,

$$L(k+1) \geq L(k) \text{ if and only if } (c_1 - pc_2)Q(k) \geq c_1 q. \tag{1}$$

From (1) it follows that $L(k+1) < L(k)$ for all k if $c_1 - pc_2 \leq 0$.

(c) The first statement is immediate from (1). If the inequality $Q(k) \geq \beta$ holds for some value of k, there is a finite optimal value of k; otherwise, the sequence $\{L(k)\}$ is strictly decreasing, and the optimal value of k is $k = \infty$.

(d) Lemma 3.2 asserts that $\lim_{k\to\infty} A(k) = (r(\infty))^{-1}$. It follows from the Monotone Convergence Theorem that $\lim_{k\to\infty} D(k) = \gamma$. Using these two facts and the second inequality, we conclude that $\lim_{k\to\infty} Q(k) > \beta$. From part (b), it follows that a finite optimal value of k* of k exists and that

$$k^* = \min\{k: Q(k)\} \geq \beta.$$

This concludes the proof of Theorem 3.3.

3.4 Remark. Recalling the definition of β in Theorem 3.3, suppose that $r(\infty) > \beta(p\mu)^{-1}$. It follows immediately that $r(\infty) > \beta(p\gamma)^{-1}$ and, hence, from Theorem 3.3 that a finite optimal value k* of k exists. It is possible to compute an upper bound k** of k*, in this special case, by a method we now discuss. The method is useful because k** is sometimes easier to compute. Let

$$k^{**} = \min\{k: (A(k))^{-1} > (p\mu)^{-1}\beta\}. \tag{2}$$

Since $\lim_{k\to\infty} (A(k))^{-1} = r(\infty)$, k** is well-defined. The inequalities

$$Q(k) > \frac{pD(1)}{A(k)} = \frac{p\mu}{A(k)} > \beta$$

for $k \geq k^{**}$ show that $k^{**} \geq k^*$.

3.5 Example. Let $F(x) = 1 - e^{-(\lambda x)^\alpha}$ for some $\alpha > 1$ and some $\lambda > 0$ and all $x \geq 0$. We find that

$$A(k) = \frac{\Gamma(k+1/\alpha)}{\lambda\alpha k!} = \frac{\Gamma(1/\alpha)}{\lambda\alpha k!} \prod_{\ell=0}^{k-1} (\ell + 1/\alpha).$$

Since the expression for D(k) is rather complex, it is impossible to give a simple expression for k*; in practice, k* must be found by evaluating Q(k) and checking whether $Q(k) \geq \beta$. We shall indicate how to find an upper bound for k**. To do this, we use the inequality

$$A(k) \leq \int_0^T e^{-R(t)} \frac{R(t)^k}{k!} dt + \frac{1}{r(t)}, \tag{3}$$

for k 1, 2, 3, ... and T > 0, which is used in the proof of Lemma 3.2 in [6]. In the present example, $R(t) = (\lambda t)^{\alpha}$. Replacing this factor in the integral in (3) by the constant 1, we get

$$A(k) \leq \frac{\lambda^{k\alpha} T^{k\alpha+1}}{(k\alpha+1)k!} + \frac{1}{\alpha\lambda^{\alpha} T^{\alpha-1}}. \tag{4}$$

Choose T so that

$$\alpha\lambda^{\alpha} T^{\alpha-1} > 2\beta(p\mu)^{-1}.$$

Then choose k so that, for this value of T,

$$\frac{\lambda^{k\alpha} T^{k\alpha+1}}{(k\alpha+1)k!} < \frac{p\mu}{2\beta}.$$

Using (2) and (4), we find that

$$k^{**} \leq k.$$

4. MODEL III

Our third model, based on a model in [6], deals with a unit subject to two types of failures. We assume that the unit has life distribution F and that when the unit fails, with probability η, the failure is a Type 1 failure and that, with probability $1-\eta$, the failure is a Type 2 failure. When the unit has a Type 1 failure, it is repaired at a cost of c_1. With probability p, this repair is perfect; otherwise, it is minimal. When the unit has a Type 2 failure or when it fails again after $k-1$ successive imperfect repairs following Type 1 failures, the unit is (perfectly) overhauled or replaced at a cost c_2 $(c_2 > c_1)$. The problem is to choose k so as to minimize L(k), the long-run expected cost per unit time. Since the proofs of the basic results for this model are much like those for Model II, we shall state our results without proof. (The derivation of D(k) is the same as in Theorem 3.1 except that p is replaced by $1-\eta q$, the probability that a failure is corrected by perfect repair.)

4.1 *Theorem.*

$$L(k) = \frac{c_1\left(\eta - \eta(\eta q)^{k-1}\right) + c_2\left(1-\eta+\eta(\eta q)^{k-1} - (\eta q)^k\right)}{(1-\eta q)\sum_{\ell=0}^{k-1}(\eta q)^{\ell}A(\ell)}$$

4.2 *Theorem.* Let

$$Q(k) = \frac{(1-\eta q)D(k)}{A(k)} + (\eta q)^k$$

where

$$D(k) = \sum_{\ell=0}^{k-1}(\eta q)^{\ell}A(\ell).$$

Also let

$$\delta = \int_0^{\infty} \bar{F}^{(1-\eta q)}(x)\,dx$$

and

$$\beta = \frac{q\left(c_1\eta + c_2(1-\eta)\right)}{c_1 - c_2(1-\eta q)}\ .$$

(a) The sequence $\left(Q(k)\right)$ is increasing.

(b) If $c_1 \leq c_2(1-\eta q)$, then the sequence $\left(L(k)\right)$ is decreasing and the optimal value of k is $k = \infty$.

(c) Suppose that $c_1 > c_2(1-\eta q)$. Then, for any positive integer k, $L(k+1) \geq L(k)$ if and only if $Q(k) \geq \beta$. Also, an optimal value of k, possibly infinity, exists.

(d) Suppose that $c_1 > c_2(1-\eta q)$ and that $r(\infty) > \beta\left(\delta(1-\eta q)\right)^{-1}$. Then a finite optimal value k^* of k exists. Furthermore,

$$k^* = \min\{k\colon Q(k) \geq \beta\}.$$

5. INCREASING REPAIR COSTS

In this section we present two modifications, based on the main cost structure considered in [2], of our first and second models.

The first of these two modified models, Model IV, is exactly like Model I except that the cost of the ℓth repair, beginning with a new or perfectly repaired unit, is $a + c\ell$ for some constants $a > 0$ and $c > 0$. We denote the cost of replacement or perfect overhaul by c_0. The objective is to find the replacement age T which minimizes L(T), the long-run expected cost per unit time.

In Theorem 5.1 and Theorem 5.2 we give a formula for L(T) and discuss minimization of L(T). Then we apply these results to the exponential distribution in Example 5.3.

5.1 Theorem.

$$L(T) = \frac{\left(\frac{a}{p}+\frac{c}{p^2}\right)G(T) + \left(c_0 - \frac{cq}{p}R(T)\right)\overline{G}(T)}{\int_0^T \overline{G}(x)\,dx} .$$

Proof. From renewal theory, we have that

$$L(T) = \frac{C(T)}{D(T)} ,$$

where C(T) and D(T) have the expected interpretations. D(T) is, of course,

$$\int_0^T \overline{G}(x)\,dx,$$

just as in Model I. For $k = 1, 2, \ldots$, let

$$c_k = \sum_{i=1}^{k} (a+ic) = \left(a+\frac{c}{2}\right)k + \frac{c}{2}k^2 .$$

Then from Lemma 2.1, we obtain that

$$C(T) = \int_0^T g(x)\left(\sum_{k=0}^{\infty} \frac{\left(qR(x)\right)^k}{k!} e^{-qR(x)} c_{k+1}\right)dx$$

$$+ \bar{G}(T)\left(c_0 + \sum_{k=1}^{\infty} \frac{(qR(T))^k}{k!} e^{-qR(T)} c_k\right) .$$

Simplifying the infinite series in the two parts of the expression, we obtain that

$$C(T) = \int_0^T g(x)\left((a + 2c)qR(x) + \frac{c}{2} q^2R^2(x) + (a + c)\right)dx$$

$$+ \bar{G}(T)\left(c_0 + (a + c)qR(T) + \frac{c}{2} q^2R^2(T)\right) .$$

Using the identities

$$\int_0^T R(x)g(x)dx = \frac{G(T)}{p} - \bar{G}(T)R(T)$$

and

$$\int_0^T R^2(x)g(x)dx = \frac{2G(T)}{p^2} - \frac{2}{p}\bar{G}(T)R(T) - \bar{G}(T)R^2(T),$$

we have, after simplification, that

$$C(T) = \left(\frac{a}{p} + \frac{c}{p^2}\right)G(T) + \left(c_0 - \frac{cq}{p} R(T)\right)\bar{G}(T),$$

from which the claimed expression for $L(T)$ follows.

5.2 *Theorem*. Assume that $r(t)$ is differentiable and nondecreasing and that $cq > 0$. For $t \geq 0$, let

$$H(t) = \int_0^t \bar{G}(x)dx\left(a + c - c_0p + cqR(t)\right)r(t) - \left(\frac{a}{p} + \frac{c}{p^2}\right)G(t)$$

$$+ \left(\frac{cq}{p} R(t) - c_0\right)\bar{G}(t).$$

(a) $L'(t) = \bar{G}(t)\left(\int_0^t \bar{G}(x)dx\right)^{-2} H(t).$

(b) For any sufficiently large number t, L is increasing on $[t, \infty)$.

(c) If $a + c > pc_0$ and either $r(0) > 0$ or r is increasing, then H is increasing. In this case, the optimal value of T is the unique solution of the equation $H(T) = 0$.

Proof. (a) This is straightforward.

(b) Note that $R(t) \to \infty$ as $t \to \infty$ since r is nondecreasing and not identically zero. Also note that $\overline{G}(t) \to 0$ as $t \to \infty$. Thus $H(t) \to \infty$ as $t \to \infty$. The conclusion then follows from part (a).

(c) Since any optimal value of T is a solution of the equation $H(T) = 0$, by part (a), it suffices to verify that H is increasing. We do this by computing $H'(t)$ to obtain

$$H'(t) = \int_0^t \overline{G}(x)dx\left[\left(a + c - c_0p + cqR(t)\right)r'(t) + cqr^2(t)\right].$$

This concludes the proof of the theorem.

5.3 Example. Let $\overline{F}(x) = e^{-\lambda x}$, for some $\lambda > 0$ and all $x \geq 0$. The optimal age T at which to replace a unit with life distribution F, assuming that $a + c > c_0p$ and $cq > 0$, is the unique solution of the equation $H(T) = 0$. This equation reduces to the equation

$$cqe^{-p\lambda T} + cqp\lambda T = c_0p^2 + cq.$$

The last model that we consider, Model V, is the same as Model II except that the cost of the ℓth repair, beginning with a new or perfectly repaired unit, is $a + c\ell$, for some constants $a > 0$ and $c > 0$. We denote the cost of replacement or overhaul by c_0. The object is to find the value of k which minimizes $L(k)$, the long-run expected cost per unit time.

In the remainder of this section, we give a formula for $L(k)$ (Theorem 5.4), discuss optimization of $L(k)$ (Theorem 5.5), and present an example involving the exponential distribution (Example 5.6).

5.4 Theorem.

$$L(k) = \frac{C(k)}{D(k)}$$

where

$$C(k) = p^{-2}\Big((c+ap) + (c_0 - ck - a)q^{k-1} + (ck - c + a - 2c_0)q^k + c_0 q^{k+1}\Big)$$

and

$$D(k) = \sum_{\ell=0}^{k-1} q^{\ell} A(\ell).$$

Proof. Use the idea used in the proof of Theorem 3.1.

5.5 Theorem. Let

$$Q(k) = \frac{\big(pD(k) + q^k A(k)\big)(pck + ap - p^2 c_0)}{A(k)} + cq^{k+1}.$$

(a) $L(k+1) \geq L(k)$ if and only if $Q(k) \geq (ap+c)q$.

(b) There is a finite optimal value k^* of k.

(c) If $a+c \geq pc_0$, then the sequence $\big(Q(k)\big)$ is increasing. In this case,

$$k^* = \min\{k: Q(k) \geq (ap+c)q\}.$$

Also, the sequence $\big(Q(k)\big)$ is eventually increasing even if $a+c < pc_0$.

Proof. (a) This is straightforward.

(b) This is immediate from (a) since $Q(k) \to \infty$ as $k \to \infty$.

(c) The inequality

$$Q(k+1) > Q(k)$$

is equivalent to the inequality

$$p^2 cA(k)D(k+1) + (pck + ap - p^2 c_0)\big(pD(k) + pq^k A(k)\big)\big(A(k) - A(k+1)\big) > 0.$$

Recall (Lemma 3.2) that the sequence $\big(A(k)\big)$ is decreasing. Thus if $ap - p^2 c_0 \geq 0$ or if k is sufficiently large, then $Q(k+1) > Q(k)$.

5.6 Example. Let $\overline{F}(x) = e^{-\lambda x}$ for some $\lambda > 0$ and all $x \geq 0$. A calculation involving the gamma function shows that

$$A(k) = \frac{1}{\lambda} \quad \text{and} \quad D(k) = \frac{1 - q^k}{p\lambda} .$$

The inequality $Q(k) \geq (ap + c)q$ is equivalent to the simple inequality

$$pck + cq^{k+1} \geq cq - p^2(a - c_0),$$

whose smallest solution is the optimal value k^*.

ACKNOWLEDGMENT

We wish to thank an anonymous referee who suggested that we introduce Model U, prove a result similar to Lemma 1.1, and then use that result in the proofs of Lemma 2.1 and Theorem 3.1. His suggestions resolved some points of confusion centering around Lemma 2.1 and its proof and, in so doing, greatly improved our paper.

REFERENCES

[1] R. Barlow and F. Proschan, *Mathematical Theory of Reliability*, John Wiley and Sons, New York, 1965.

[2] P. Boland and F. Proschan, Periodic replacement with increasing minimal repair costs at failure, *Op. Res. 30* (1982), pp. 1183-1189.

[3] M. Brown and F. Proschan, Imperfect repair, to appear in *J. Appl. Prob.*

[4] M. Brown and F. Proschan, Imperfect maintenance, *IMS Lecture Notes*, Volume 2, 1983, pp. 179-188.

[5] T. Nakagawa, A summary of imperfect maintenance policies with minimal repair, *R.A.I.R.O 14* (1980), pp. 249-255.

[6] T. Nakagawa, Generalized models for determining optimal number of minimal repairs before replacement, *J. Op. Res. Soc. of Japan 24* (1981), pp. 325-337.

OPTIMAL REPLACEMENT FOR SELF-REPAIRING SHOCK MODELS

Gary Gottlieb

Graduate School of Business Administration
New York University
New York, New York

Benny Levikson

Department of Statistics
Haifa University
Haifa, Israel

This paper contains a summary of Gottlieb and Levikson (1984) which was the basis of their talk at the conference. Other closely related work by Taylor (1975), Feldman (1976) and Gottlieb (1982) is reviewed to give some perspective to the problem being considered and to provide a brief introduction to the area of optimal replacement. No new results are presented.

In all the papers considered, a device is subject to a series of shocks, each of which increase the cumulative damage Z(t) incurred by the device and each of which may cause the device to fail. Let ξ be the failure time of the device and T the replacement time. A cost $C_0 > 0$ is incurred each time the device is

ISBN 0-12-041420-1

replaced (replacement is mandatory upon failure) and an additional cost $C > 0$ is incurred if $T \geq \xi$.

Let $P_x(\cdot) = P(\cdot | Z(0) = x)$ and $E_x(\cdot) = E(\cdot | Z(0) = x)$ where $P(\cdot) = P_0(\cdot)$ and $E(\cdot) = E_0(\cdot)$. Let $\hat{Z}(t) = Z(t)$ if $t < \xi$ and $\hat{Z}(t) = \Delta$ if $t \geq \xi$ where $\Delta \notin \mathbb{R}$. The objective is to find the T in the specified class of replacement times which minimizes

$$\psi_T = \frac{C_0 + CP(\xi \leq T)}{E(T \wedge \xi)} .$$

Here $a \wedge b = \min(a,b)$. Henceforth, let $C_0 = 1$ without loss of generality.

Taylor's paper can be viewed as the seminal work. He takes the damage process to be a compound Poisson Process with positive jump sizes and makes the universal assumption that failure can occur only at jump times. He assumes that the probability a jump causes a functioning device to fail is an increasing function of the cumulative damage after the jump. This implies an increasing failure rate in time over all sample paths and Taylor shows that over the class of all Markovian replacement policies, a control limit policy is optimal. He indicates how to calculate this limit.

Feldman uses a Semi-Markov Process to model the damage process. He assumes an increasing failure rate and restricts his attention to control limit policies. His main concern is how to determine the optimum control limit.

In Gottlieb, a Semi-Markov Process is also used to model the damage process. The failure rate is not assumed to be increasing and all Markovian replacement policies are considered. Using Markov Decision Theory, conditions are established where a control limit policy is optimal and conditions are given where the optimal policy has a different but useful form.

In Feldman and Joo (1984), the best algorithm to date is given to compute the optimal replacement policy for a special case of the problem studied by Gottlieb.

In Gottlieb and Levikson, a model which allows for repairs between shocks is given. This model generalizes previous work, while the results are similar to those of Gottlieb.

In Taylor, shocks arrive as a Poisson process with rate λ. Each shock increases the total damage by a positive amount distributed according to F. The system survives a shock which increases the damage to x with probability $r(x)$, r a decreasing function. Taylor shows that the T which minimizes ψ_T over the class of replacements is

$$T^* = \inf\{t \geq 0 : Z(t) \geq x^*\}$$

Taylor shows that x^* and $\psi^* = \inf_T \psi_T$ can be determined by noting that

$$x^* = \inf\{x \geq 0 : \psi^* - \lambda C(1 - R(x)) \leq 0\}.$$

Here, $R(x) = \int r(x + y)\, dF(y)$ and $\lambda(1 - R(x))$ is the failure rate at time t if $Z(t) = x$.

To derive this result, note that a Markov replacement policy T minimizes ψ_T if and only if it maximizes

$$\rho_T = \psi^* E(T \wedge \xi) - CP(T \geq \xi)$$

and that the maximum value of ρ_T is 1. Now,

$$E(T \wedge \xi) = E(\xi) - E(\xi - T; T < \xi).$$

Also,

$$E(\xi - T \wedge \xi) = E_{Z(T \wedge \xi)}(\xi).$$

Letting

$$w(x) = E_x(\xi),$$

$$\Phi(x) = \psi^* w(x) - C,$$

$$\rho_T = -(E(\Phi(X(T));T < \xi) - \Phi(0))$$

Recall that $w(x)$ satisfies $Gw(x) = -1$ where G is the infinitesimal operator for the Markov process $\hat{Z}(t)$. Let $f : \mathbf{R} \cup \{\Delta\} \to \mathbf{R}$ with $f(\Delta) = 0$. Then, for $x \in \mathbf{R}$,

$$\begin{aligned} Gf(x) &= \lim_{h\downarrow 0} h^{-1} E_x\{f(Z(h)) - f(x); \xi > h\} \\ &= \lim_{h\downarrow 0} h^{-1}\{(1 - \lambda h)f(x) + \lambda h \int r(x + y)\, dF(y) - f(x)\} \\ &= -\lambda\{f(x) - \int r(x + y) f(x + y)\, dF(y)\}. \end{aligned}$$

Using Dynkin's formula,

$$\begin{aligned} E_x(f(\hat{Z}(T))) &= E_x(f(Z(T)); T < \xi) \\ &= f(x) + E_x \int_0^T Gf(Z(u))du. \end{aligned}$$

So,

$$\rho_T = -E\int_0^T G\Phi(Z(u))du.$$

As $G\Phi(x) = -\psi^* - \lambda C(1 - R(x))$ and R is monotone and $Z(t)$ is an increasing process, Taylor concludes that x^* is the smallest value of $x \geq 0$ for which $-G\Phi(x) \leq 0$.

Feldman models his damage process by a right-continuous nondecreasing semi-Markov process $Z(t)$ with state space $\mathbb{R}^+$ and associated Markov renewal process (X_n, β_n) where β_n is the time of the nth transition and $X_n = Z(\beta_n)$. He models ξ exactly as does Taylor. Let $\hat{X}_n = \hat{Z}(\beta_n)$.

He defines

$$Q(x,y,t) = P(X_{n+1} \in [x,y],\ \beta_{n+1} - \beta_n \leq t | X_n = x),$$

$$\hat{Q}(x,y,t) = P(\hat{X}_{n+1} \in [x,y],\ \beta_{n+1} - \beta_n \leq t | \hat{X}_n = x).$$

So,

$$\hat{Q}(x,y,z) = \int Q(x,du,t)r(u)du.$$

Feldman only considers replacement policies of the form

$$T_a = \inf\{t \geq 0 : Z(t) \geq a\}.$$

Letting

$$R(x,y,t) = \sum_{n=0}^{\infty} Q^n(x,y,t),$$

where the convolutions have their usual meaning,

$$q(u) = 1 - Q(u,\infty,\infty),$$

$$m(u) = \int_0^{\infty} (1 - \hat{Q}(u,\infty,t))dt,$$

he shows that

$$\psi_{T_a} = \frac{1 + CP(T_a \geq \xi)}{E(T_a \wedge \xi)} = \frac{1 + C\int_0^a R(x,du,\infty)q(u)}{\int_0^a R(x,du,\infty)m(u)} .$$

Using the above formula, Feldman showed that if $E_u(\beta_1)$ is decreasing in u and $P_u(\hat{X}_1 = \Delta)$ is increasing in u, the optimal replacement level x^* is the unique solution to

$$\frac{1}{m(x)} \int R(x,du,\infty)(q(x)m(u) - m(x)q(u)) = C^{-1}.$$

If no solution exists, $T^* = \infty$ is optimal.

Gottlieb considers the same model as Feldman. He allows for replacement at all points in time. He does not assume an increasing failure rate.

He studies the equivalent Markov Decision problem based on the process X_n, action space $A = \{t : t \geq 0\}$ where choosing action t at time β_n means replace at time $\beta_n + t$ if $\beta_{n+1} - \beta_n > t$, and with transition probabilities for the induced Markov chain given below:

$$P(\hat{X}_{n+1} \in [x,z] \mid \hat{X}_n = x, \text{ action } t) = G_x(t)\int_0^{z-x} r(x+y)\, dF_x(y)$$

$$P(\hat{X}_{n+1} = \Delta \mid \hat{X}_n = x, \text{ action } t) = 1 - G_x(t) + G_x(t)\int (1 - r(x+y))\, dF_x(y)$$

$$P(\hat{X}_{n+1} = \Delta \mid \hat{X}_n = x, \text{ any action}) = 1.$$

Here, G_x is the distribution of the sojourn time in x and F_x is the distribution of the jump size out of state x.

Of course, these probabilities are also given by the appropriate $\hat{Q}$'s in Feldman's notation.

Finally, if action t is chosen when in state $x \in \mathbb{R}$, then a reward

$$C(x,t) = \psi^* \int_0^t s\, dG_x(s) + t(1 - G_x(t))\psi^* - C(1 - R(x))G_x(t)$$

is received. Also, the reward given when in state Δ is 0.

Define

$$V(x) = \sup_{T \text{ Markovian}} \psi^* E_x(T \wedge \xi) - CP_x(T \leq \xi), \quad x \in \mathbb{R},$$

$$V(\Delta) = 0 .$$

Of course, $V(0) = 1$ by previous arguments.

Gottlieb argues that $V(x)$ is the unique solution to $TV(x) = V(x)$ where T is the monotone contraction operator defined by:

$$Tu(x) = \begin{cases} \sup_{t \geq 0} \{C(x,t) + G_x(t)\int r(x+y)u(x+y)\, dF_x(y)\} & \text{if } x \in \mathbb{R} \\ 0 & \text{if } x = \Delta. \end{cases}$$

Further, he argues that the optimal action if $\hat{X}_n = x$ is $t^*(x)$ where t^* is the smallest t satisfying

$$V(x) = C(x,t^*) + G_x(t^*)\int_0^\infty r(x+y)V(x+y)\, dF_x(y) .$$

Knowing $t^*(x)$, it is a simple matter to define T^*.

Let $h(x,u) = \frac{\partial G_x(u)}{\partial u} / (1 - G_x(u))$. $h(x,u)$ is the hazard rate of G_x at u. Gottlieb assumes that $r(x)$ is decreasing in x and that $x + Y_x$ is stochastically increasing in x where $Y_x \sim F_x$. His major result is as follows:

Assume that $h(u,x)$ is increasing in x for each u. Then,

(1) If $h(x,u)$ is increasing in u for each x, $t^*(x)$ is decreasing in x.

(2) If $h(x,u)$ is decreasing in u for each x, a control limit policy is optimal. That is, for some x^* with $0 \leq x^* \leq \infty$,

$$t^* = \begin{cases} \infty \text{ if } x < x^*. \\ 0 \text{ if } x \geq x^*. \end{cases}$$

So, in this case,

$$T^* = \inf\{t \geq 0 : Z(t) \geq x^*\}.$$

In case (1), $T^* = \inf\{t \geq 0 : t \in [\beta_n, \beta_{n+1}]$, some n and $t - \beta_n > \phi(X_n)\}$ where ϕ is a decreasing function.

We now summarize the results of Gottlieb and Levikson.

They assume that the cumulative damage is a decreasing Markov Process except for a sequence of strictly positive jumps (shocks) which arrive at a rate λ which is a function of the present state and the time since the last jump and where the distribution of the jump size depends only on the present state. A semi-Markov Process is a special case of this process. They assume that failure of the device can occur only at times of jump and that the probability that a shock will cause a functioning device to fail is a function of the cumulative damage after the shock.

They show that if the jump rate is an increasing function of the state and a decreasing function of the time since the last jump, then subject to some further natural conditions, the optimal replacement policy is to replace as soon as the cumulative damage equals or exceeds some threshold level (control limit policy).

If the jump rate is an increasing function of both the state and the time since the last jump, then subject to some further natural conditions, the optimal replacement policy is to replace as soon as the time since the last jump equals or exceeds some level which is a decreasing function of the cumulative damage.

We refer the reader to Gottlieb and Levikson (1984) for the proofs of the theorems and additional results.

The damage process $\{Z(t), t \geq 0\}$ is decomposable into an ordered pair $\{(X(t), S(t)), t \geq 0\}$ where $S(t)$ is the time since the last jump and $X(t)$ is the present value of total cumulative damage and is taken to be right-continuous.

We assume that $X(t)$ is a decreasing, time-homogeneous Markov process, as long as no shock occurs. The decreasing property is imposed to describe the fact that maintenance work is being done. Shocks arrive randomly but depending on the time $S(t)$ since the last shock and the present value of the cumulative damage. Thus in the time interval $(t, t + \Delta t)$ the probability of a shock is $\lambda(X(t), S(t))\Delta t + o(\Delta t)$. When a shock occurs its magnitude depends on the present cumulative damage x. That is, the shock magnitude $Y_x \sim F_x$. Thus the cumulative damage, immediately after a shock at time t, is $X(t) = X(t-) + Y_{X(t-)}$. The probability of surviving such a shock is given by $r(X_t)$. Letting ξ = failure time, let $\hat{X}(t) = X(t)$ for $t < \xi$ and $\hat{X}(t) = \Delta$ for $t \geq x$. Next we give the assumptions imposed on the various random phenomena described above.

<u>ASSUMPTION 1</u>. If $Y_x \sim F_x$, then $x + Y_x$ is stochastically increasing in x.

<u>ASSUMPTION 2</u>. For $t \geq 0$, $x \in \mathbb{R}$, $0 < \lambda_0 = \lambda(x,t) = \Lambda_0 < \infty$.

<u>ASSUMPTION 3</u>. Let L be the infinitesimal generator of a decreasing time-homogeneous Markov Process with the assumption that if two versions of the process evolve simultaneously, with one beginning at x and the other at $y > x$, then there is a probability space so that on this space, if the second ever gets below the first, they must have been equal at some earlier time.

ASSUMPTION 4. Let $\{Z(t), t \geq 0\}$ be a Markov Process with infinitesimal generator H defined as follows. For all ζ in the domain of H, let

$$H\phi(x,t) = L\zeta(x,t) + \frac{\partial}{\partial t}\zeta(x,t)$$

$$+ \lambda(x,t)\int(\zeta(x+y,0) - \zeta(x,t))\, dF_x(y).$$

Here, $\lambda(x,t)$ is the jump rate and F_x is the jump-size distribution out of state x.

ASSUMPTION 5. Let r(x) be the probability that the device will survive a shock which beings the cumulative damage to a level x. We assume that r(x) decreases in x and let

$$R(x) = \int_{y \geq 0} r(x+y)\, dF_x(y),$$

$$\bar{R}(x) = 1 - R(x).$$

We note that for Assumptions 2 and 5, $\bar{R}(x)$ is decreasing in x.

ASSUMPTION 6. Let $\beta_0 = 0$,

$$\beta_1 = \inf\{t > 0 : S(t) = 0\},$$

$$\vdots$$

$$\beta_n = \inf\{t > \beta_{n-1} : S(t) = 0\}.$$

β_n is the time of the nth jump or shock. Let ξ be the lifetime of the device. We assume that

$$P(\xi > t \mid Z(u),\ u \leq t,\ \beta_n \leq t < \beta_{n+1}) = \prod_1^n r(X(\beta_i)).$$

Note that the failure rate at time t is $\lambda(X(t),S(t))\bar{R}(X(t))$ and that

$$\xi \quad \{\beta_n\}_0^\infty \cup \{\infty\}.$$

Let

$$P_{x,t}(\cdot) = P(\cdot|X(0) = x, S(0) = t),$$

$$E_{x,t}(\cdot) = E(\cdot|X(0) = x, S(0) = t).$$

We will assume without loss of generality that $E_{x,t}(\xi) < \infty$, all x and t. Let $\mathcal{C}$ be the set of all stopping times with respect to the process $\{Z(t), t \geq 0\}$.

For $T \in \mathcal{C}$, define

$$\psi_T = \frac{1 + CP_{0,0}(\xi \leq T)}{E_{0,0}(\xi \wedge T)}.$$

As before, let $\psi^* = \inf_{T \in \mathcal{C}} \psi_T$ and $\rho_T = \psi^* E_{0,0}(\xi \wedge T) - CP_{0,0}(\xi \leq T)$. As before, for $T^* \in \mathcal{C}$, $\psi_{T^*} = \psi^*$ if and only if $\rho_{T^*} = \sup_{T \in \mathcal{C}} \rho_T$. Let

(1) $$V(x,t) = \sup_{T \in \mathcal{C}} \psi^* E_{x,t}(\xi \wedge T) - CP_{x,t}(\xi \leq T).$$

It is convenient to study ρ_T and more generally, V(x,t), rather than ψ_T.

The properties of V(x,t) and of the optimal Replacement Policy are summarized below. Throughout, we assume that Assumptions 1-6 hold and we will specify which if any of the following two assumptions hold.

ASSUMPTION 7. Let $\lambda(x,t)$ be increasing in x and decreasing in t.

ASSUMPTION 8. Let $\lambda(x,t)$ be increasing in x and increasing in t.

As mixtures of DFR distributions are DFR (see Barlow and Proschan, 1975, p. 103), Assumption 7 implies that the time between shocks has a DFR distribution while Assumption 8 does not in general imply that the time between shocks has an IFR distribution.

The two major results are as follows:

THEOREM 1. Under Assumption 7, V(x,t) is decreasing in x and increasing in t.

THEOREM 2. Under Assumption 8, $V(x,t)$ is decreasing in x and in t.

Theorems 1 and 2 are used to derive results about the optimal replacement policy. Theorem 3.3, p. 127 of Shiryayev (1978) considers the problem of stopping a Markov process in order to maximize a terminal reward which is a function of the state of the process when it is stopped. We can put (1) into this framework by expanding the state space of $Z(t)$ by adjoining the ordered pair (t,M_t) where t is the time index and $M_t = \xi$ if $\xi \le t$ and $M_t = -1$ if $\xi > t$, Let the terminal reward at time t be $\psi^* t$ if $M_t = -1$ $(\xi > t)$ and $\psi^* \xi - c$ if $\xi \le t$. Given this terminal reward, the problem is a special case of the problem Shiryayev considered.

It is a simple matter to show that the conditions of Theorem 3.3, part 3 of Shiryayev are satisfied by our problem and it then follows that an optimal replacement policy T^* exists and satisfies

$$T^* = \inf\{t \ge 0 : V(X(t),S(t)) = 0\}$$

From the above and Theorems 1 and 2, we immediately get the following results.

THEOREM 3. If Assumption 7 holds, then

$$T^* = \inf\{t \ge 0 : X(t) \ge x^*\}$$

where

$$x^* = \inf\{x \ge 0 : V(x,0) = 0\}.$$

THEOREM 4. If Assumption 8 holds, then there exists a decreasing function $Q : \mathbf{R} \to \mathbf{R}^+$ so that

$$T^* = \inf\{t \ge 0 : S(t) \ge Q(X(t))\}$$

where

$$Q(x) = \inf\{t \ge 0 : V(x,t) = 0\}.$$

The policy in Theorem 3 is a control limit policy.

THEOREM 5. If assumption 7 holds and if $\{X(\beta_n)\}_{n=0}^{\infty}$ is an increasing process, then $x^* = y^*$, where $y^* = \inf\{x \geq 0 : \psi^* a(x) - - Ch(x) \leq 0\}$ where

$$a(x) = E_x(\beta_1),$$

$$b(x) = P_x(\beta_1 = \xi).$$

REFERENCES

Barlow, R., and Proschan, F., 1975. *Statistical Theory of Reliability and Life Testing Probability Model.* Holt, Rinehart and Winston, New York.

Bergman, B., 1978. Optimal Replacement Under a General Failure Model. *Adv. Appl. Prob. 10,* 431-451.

Feldman, R. M., 1976. Optimal Replacement with Semi-Markov Shock Models. *J. Appl. Prob. 13,* 108-117.

Feldman, R. M., and Joo, N. Y. *A State-Age Dependent Policy of a Shock Process* (forthcoming).

Gottlieb, G., 1982. Optimal Replacement for Shock Models with General Failure Rate. *Opers. Res. 30,* 82-92.

Gottlieb, G., and Levikson, B., 1984. Optimal Replacement for Self-Repairing Shock Models with General Failure Rate. *J. Appl. Prob.* (forthcoming March).

Mine, H., and Osaki, S., 1970. *Markovian Decision Processes.* American Elsevier, New York.

Shiryayev, A. N., 1978. *Optimal Stopping Rules.* Springer-Verlag, New York.

Taylor, H. M., 1975. Optimal Replacement Under Additive Damage and Other Failure Models. *Naval Res. Logist. Quart. 22,* 1-18.

Zuckerman, D., 1978. Optimal Stopping in a Semi-Markov-Shock Model. *J. Appl. Prob. 15,* 629-634.

AN ITERATIVE SCHEME FOR APPROXIMATING OPTIMAL REPLACEMENT POLICIES

Joseph Quinn[1]

Department of Mathematics
University of North Carolina at Charlotte
Charlotte, North Carolina

A system subject to wear and/or damage is modeled by a stochastic process X_t. *A replacement policy is a Markov time which determines system replacement in terms of the wear/damage history. An* ϵ*-optimal replacement policy* τ_ϵ *is one such that* $Eg(X_{\tau_\epsilon})/E(\tau_\epsilon)$ *is within* ϵ *of the minimum long-run average cost. In this paper, an iterative scheme for obtaining approximately optimal replacement policies is developed. The implementation of the scheme for certain Markovian wear/damage models is discussed and is implemented for exponentially distributed compound Poisson processes.*

[1]*Research supported by the Air Force Office of Scientific Research under Grants AFOSR-80-0245B and 0245C.*

ISBN 0-12-041420-1

I. INTRODUCTION

The wear of a device is given by a nonnegative, real valued, nondecreasing process X_t. Failure of the device is governed by a random function on the path space. At any time t prior to failure the device can be replaced at a cost $g(X_t)$. At failure the device is replaced at a cost K which will usually include some penalty amount. The optimal replacement problem is to determine the minimum long-run average cost and, if possible, to find an optimal policy (or, more practically, a nearly optimal policy). In this paper we develop an iterative scheme which, for a wide class of models, can be used to provide both approximately optimal policies and approximations to the minimal long-run average cost.

Early work on this kind of problem emphasized age replacement policies, c.f. [5], chapter 3 and [9], chapter 11 for a review of much of this work. In these models the age of the system is the only control variable. Replacement upon failure or first reaching a fixed critical age has been shown to be an optimal policy and a method for determining the critical age is known. Work on these kinds of age models continues with emphasis on repair and maintenance policies, c.f. [14].

The problem of finding optimal policies for wear processes has been studied by J.M. Taylor [19], B. Bergman [6], R.M. Feldman [11], [12] and [13], O. Zuckerman [21] and [22], G. Gottlieb [15], M. Abdel-Hameed [1] and [2], Abdel-Hameed and F. Proschan [3], and Abdel-Hameed and I.N. Shimi [4]. All of these works seek true optimal policies, which don't exist in sufficiently general settings. Furthermore, with the exception of [2], the cost function g is assumed to be constant. This along with other assumptions has usually permitted the optimal policy to be characterized as a control limit policy and sometimes a method for computing the control limit has also been provided (see in particular [19] and [22]). In [15] more general failure rates are considered which result in the determination of some non-control-limit optimal policies. As indicated before, the only prior consideration of non-constant cost functions is in [2]. Here the costs can be arbitrary positive, bounded functions of the state but a solution to the optimal replacement problem is provided only under conditions which are strong enough to produce a control limit policy. No method for computing the control limit is given there.

Most recently, G. Gottlieb and B. Levikson [16] have considered optimal replacement for self-repairing models. Here the wear process is no longer required to be nondecreasing (the cost function is still constant).

The iterative procedure developed in this paper is powerful enough to provide approximately optimal policies for a superset of all the Markovian models considered so far in the literature. Only very reasonable and mild assumptions are made on the cost functions and very general failure modes can be accommodated. The procedure applies to multivariate wear processes as well which in particular allows the incorporation of the time variable into the state space. This permits application to nonhomogeneous problems as well as allowing the cost function to be time dependent.

Section 2 of this paper contains definitions and other preliminaries. In Section 3 the general iterative procedure is described and justified. The results of Section 3 are then specialized, in Section 4, to the Markovian, one dimensional setting and an application to compound Poisson processes with exponentially distributed wear increments is given. This application requires discretization and is implemented by a computer program. The errors introduced by the discretization are fully discussed.

II. PRELIMINARIES

Let $X = (\Omega, X_t, F_t, F, P)$ be a stochastic process with state space (E, E). Here (Ω, F, P) is a probability space, $t \in [0,\infty)$ and $\{F_t\}$ is a filtration, i.e., each F_t is a sub-σ-algebra of F containing $\sigma(X_t)$ and satisfying $F_{t'} \subseteq F_t$ whenever $t' \le t$. The state space E is a measurable space with σ-algebra E. Let λ be an $\{F_t\}$-Markov time such that $E(\lambda) < \infty$ and let T be the set of all $\{F_t\}$-Markov times τ such that $\tau \le \lambda$.

For X, λ as above we form the "killed" process $\tilde{X}$ with state space $E^{\Delta} = E \cup \{\Delta\}$ by defining

$$\tilde{X}_t = \begin{cases} X_t & \text{if } t < \lambda \\ \Delta & \text{if } t \ge \lambda \end{cases}$$

By $B^\circ = B^\circ(X)$ we denote the class of E^{Δ}-measurable, nonnegative real valued functions g on E^{Δ} which are bounded strictly away from 0 (here E^{Δ} is the σ-algebra generated by E and $\{\Delta\}$).

For $g \in B^\circ$ and $\tau \in T$ the expected value of $g(X_\tau)$ with respect to $\tilde{X}$ is given by

$$\tilde{E}g(\tilde{X}_\tau) = E[g(X_\tau) I_{\{\tau<\lambda\}}] + g(\Delta) P\{\tau \ge \lambda\}$$

and $\tilde{E}(\tau) = E(\tau)$.

The generalized optimal replacement problem (X, λ, g) where $g \in B^\circ$ is:

(a) minimize $\tilde{E}g(\tilde{X}_\tau)/\tilde{E}(\tau)$ for $\tau \in T$.

(b) if possible, find an optimal $\tau_0 \in T$, (called an optimal policy).

This problem has been solved only in very special circumstances and even then the solution is not trivial. In this paper we are interested in the related and more practical problem of finding approximately optimal policies.

Let $\psi^* = \inf_{\tau\in T} \tilde{E}g(\tilde{X}_\tau)/\tilde{E}(\tau)$ be the optimal average cost. A policy $\tau_\epsilon \in T$ is said to be an ϵ*-optimal replacement policy* associated with the problem (X,λ,g) provided $\tilde{E}g(\tilde{X}_{\tau_\epsilon})/\tilde{E}(\tau_\epsilon) \leq \psi^* + \epsilon$. The problem of finding an ϵ-optimal policy for the generalized problem (X,λ,g) will be called the *generalized* ϵ*-optimal replacement problem*. When the process X is a wear process, i.e., is nonnegative real valued and nondecreasing, the first problem is known as the optimal replacement problem. In analogy the second could be called the ϵ-optimal replacement problem. The results we obtain for the generalized ϵ-optimal replacement problem will hold for multivariate wear processes as well as the one dimensional wear processes which have been considered in the literature so far.

It is clear that the problem of minimizing $\tilde{E}g(\tilde{X}_\tau)/\tilde{E}(\tau)$ over $\tau \leq \lambda$ is equivalent to the problem of finding a $b > 0$ such that

$$\sup_{\tau\leq\lambda} [b\tilde{E}(\tau) - \tilde{E}g(\tilde{X}_\tau)] = 0 .$$

For fixed b we will refer to the problem of maximizing $b(\tilde{E}(\tau)-\tilde{E}g(\tilde{X}_\tau)$ over $\tau \leq \lambda$ as the *b-maximization problem*. If, for the time τ', $b\tilde{E}(\tau') - \tilde{E}g(\tilde{X}_{\tau'})$ is within ϵ of the maximum value of $b\tilde{E}(\tau) - \tilde{E}g(\tilde{X}_\tau)$, we will refer to τ' as an ϵ-optimal time for the b-maximization problem.

The next results, which are needed in Section 4, enable us to compare expected values computed with respect to a given Markov chain to those computed by an approximating Markov chain.

Let $Y = (\Omega, Y_n, F_n, F, P_Y^x, \phi_n)$ be a time homogeneous, non-terminating, Markov chain with state space (E,E). The transition probability $P_Y(x,\circ)$ for Y is given by $P_Y(x,A) = P_Y^x(Y_1 \in A)$. We are adopting here the terminology of [7]. We assume that E has a unique absorbing point Δ and let $\lambda = \inf\{n | Y_n = \Delta\}$. We also assume that $P^x_Y(\lambda<\infty) = 1$ for all $x \in E$. For each n, $F_n = \sigma(X_k, 0 \leq k \leq n)$.

Let S_Y be the class of all Markov chains $Z = (\Omega, Y_n, F_n, F, P^x_Z, \phi_n)$ with state space (E,E) and with Δ as a unique absorbing state. Notice that the only differences occur in the P^x_Z's. We assume that all the processes in S_Y have the same initial distributions. For $Z_1, Z_2 \in S_Y$ define $d(Z_1,Z_2) = \sup_{x\in E} \sup_{x \in E} |P_1(x,A) - P_2(x,A)|$.

The following result is a special case of a result proved in [17]. For real valued F-measurable G define $||G|| = \sup_{\omega\in\Omega}|G(\omega)|$.

(2.1) *Theorem*. Let $Z \in S_Y$ be such that $d(Z,Y) < \epsilon$. Let G be real valued, nonnegative F_n-measurable and such that $||G|| < \infty$, then for each $x \in E$, $|E^x_Z G - E^x_Y G| \leq n\epsilon ||G||$.

(2.2) *Corollary*. If N is such that $P^x_Y(\lambda>N) < \epsilon$, there exists a $\gamma > 0$ such that $P^x_Z(\lambda>N) < \epsilon$ whenever $d(Z,Y) < \gamma$.

Proof. Let $\alpha = \epsilon - P^x_Y(\lambda>N)$. By virtue of (2.1) and keeping in mind that $\{\lambda>N\} \in F_N$, $d(Y,Z) < \alpha/N$ implies that

$|P^x_Z(\lambda>N) - P^x_Y(\lambda>N)| \leq N\alpha/N = \alpha$. Hence, $P^x_Z(\lambda>N) < \epsilon$ whenever $d(Y,Z) < \alpha/N$. The result is established.

(2.3) *Theorem*. Let $g \in bE$ and let $\epsilon > 0$. Then there exists a $\gamma > 0$ such that $|E^x_Y g(Y_\tau) - E^x_Z g(Y_\tau)| \leq \epsilon$ whenever $d(Z,Y) < \gamma$ and $\tau \leq \lambda$ is any Markov time.

Proof. Without loss of generality, assume that $||g|| \neq 0$. Choose N so that $P^x_Y(\lambda>N) < \epsilon/(3||g||)$. By virtue of (2.2), choose $\gamma_1 > 0$ such that $P^x_Z(\lambda>N) < \epsilon/(3||g||)$ whenever $Z \in S_Y$ is such that $d(Y,Z) < \gamma_1$. Let $\gamma = \min\{\gamma_1, \epsilon/(6N||g||)\}$. Then if

$d(Y,Z) < \gamma$ and τ is a Markov time with $\tau \leq \lambda$,

$$E^x_Y g(Y_\tau) = E^x_Y g(Y_\tau) I_{\{\tau\leq N\}} + E^x_Y g(Y_\tau) I_{\{\tau>N\}}$$

$$\leq E^x_Y g(Y_\tau) I_{\{\tau\leq N\}} + \epsilon/3 .$$

Applying (2.1) to the positive and negative parts of $g(Y_\tau)I_{\{\tau\leq N\}}$ we get that this last expression is less than or equal to

$$E_Z^x g(Y_\tau)I_{\{\tau\leq N\}} + \epsilon/3 + \epsilon/3$$

$$\leq E_Z^x g(Y_\tau)I_{\{\tau\leq N\}} + E_Z^x g(Y_\tau)I_{\{\tau>N\}} + \epsilon$$

$$= E_Z^x g(Y_\tau) + \epsilon \ .$$

The same argument with the roles of Y and Z reversed establishes that $E_Z^x g(Y_\tau) \leq E_Y^x g(Y_\tau) + \epsilon$ whenever $d(Z,Y) < \gamma$. But then $|E_Z^x g(Y_\tau) - E_Y^x g(Y_\tau)| \leq \epsilon$ and the result is established.

III. THE ITERATIVE METHOD

In this section we will justify the iterative method described in (3.10). For the most part, this will be done in a sequence of lemmas which follow. Throughout this section X, λ and B^o are as described in the preceding section. We let $g \in B^o$ be fixed and let $\alpha = \inf g > 0$. Finally, denote by J the quantity $\tilde{E}g(\tilde{X}_\lambda)/\tilde{E}(\lambda)$, and recall that $\psi^* = \inf_{\tau\in T} \tilde{E}g(\tilde{X}_\tau)/\tilde{E}(\tau)$.

(3.1) *Lemma*. If τ is a 1-optimal policy for the generalized replacement problem (X,λ,g), then $\tilde{E}(\tau) \geq \alpha/(J+1)$.

Proof. Let τ be a 1-optimal policy. Then $\tilde{E}g(\tilde{X}_\tau)/\tilde{E}(\tau) \leq \psi^* + 1$. Thus,

$$\alpha/(J+1) \leq \tilde{E}g(\tilde{X}_\tau)/(\psi^*+1) \leq \tilde{E}(\tau) \ ,$$

which establishes the result.

(3.2) *Lemma.* A necessary and sufficient condition that $\psi_1 > \psi^*$ is that $\sup_{\tau \le \lambda} [\psi_1 \tilde{E}(\tau) - \tilde{E}g(\tilde{X}_\tau)] > 0$

Proof. Suppose $\sup_{\tau \le \lambda} [\psi_1 \tilde{E}(\tau) - \tilde{E}g(\tilde{X}_\tau)] > 0$ and let $\tau_1 \le \lambda$ be such that $\psi_1 \tilde{E}(\tau_1) - \tilde{E}g(\tilde{X}_{\tau_1}) = \beta > 0$. Then $(\psi_1 - \psi^*)\tilde{E}(\tau_1) + \psi^* \tilde{E}(\tau_1) - \tilde{E}g(\tilde{X}_{\tau_1}) = \beta$. Since $\psi^* \tilde{E}(\tau_1) - \tilde{E}g(\tilde{X}_{\tau_1}) \le 0$ we must have that $\psi_1 - \psi^* > 0$.

For the other direction, let $\psi_1 > \psi^*$ and let τ_n be a sequence of $1/n$-optimal policies for the replacement problem (X, λ, g). Then

$$\begin{aligned} &\psi_1 \tilde{E}(\tau_n) - \tilde{E}g(\tilde{X}_{\tau_n}) \\ &= (\psi_1 - \psi^*)\tilde{E}(\tau_n) + \psi^* \tilde{E}(\tau_n) - \tilde{E}g(\tilde{X}_{\tau_n}) . \end{aligned}$$

Since

$$\tilde{E}g(\tilde{X}_{\tau_n})/\tilde{E}(\tau_n) \le \psi^* + 1/n \quad ,$$

we have that

$$\begin{aligned} &(\psi_1 - \psi^*)\tilde{E}(\tau_n) + \psi^* \tilde{E}(\tau_n) - \tilde{E}g(\tilde{X}_{\tau_n}) \\ &\ge (\psi_1 - \psi^*)\tilde{E}(\tau_n) - 1/n\tilde{E}(\tau_n) = [\psi_1 - \psi^* - 1/n]\tilde{E}(\tau_n) . \end{aligned}$$

By (3.1) this last expression is bounded below by $[\psi_1 - \psi^* - 1/n]\alpha/(J+1)$. Thus, to see that $\sup_{\tau \le \lambda} [\psi_1 \tilde{E}(\tau) - \tilde{E}g(\tilde{X}_\tau)] > 0$ we need only take n so that $1/n < \psi_1 - \psi^*$.

(3.3) *Lemma.* If $\psi^* \le \psi_1 \le J$ and $\sup_{\tau\le\lambda} [\psi_1\tilde{E}(\tau)-\tilde{E}g(\tilde{X}_\tau)] \le \epsilon$, then $0 \le \psi_1 - \psi^* \le \varepsilon(J+1)/\alpha$.

Proof. Let τ_n be 1/n-optimal for the problem (X,λ,g). Since $\sup_{\tau\le\lambda} [\psi_1\tilde{E}(\tau) - \tilde{E}g(\tilde{X}_\tau)]\le\epsilon$ we have that $\psi_1\tilde{E}(\tau_n)-\tilde{E}g(\tilde{X}_{\tau_n}) \le \epsilon$ for all $n = 1,2,\ldots,$. But then

$$\epsilon \ge \tilde{E}(\tau_n)\psi_1 - \tilde{E}g(\tilde{X}_{\tau_n})$$

$$= \tilde{E}(\tau_n)(\psi_1-\psi^*) + \tilde{E}(\tau_n)\psi^* - \tilde{E}g(\tilde{X}_{\tau_n})$$

$$\ge \tilde{E}(\tau_n)(\psi_1-\psi^*) .$$

Thus, $\psi_1-\psi^*\le \epsilon/\tilde{E}(\tau_n) \le \epsilon(J+1)/\alpha$, where the last inequality is by (3.1). The lemma is proved.

(3.4) *Lemma.* Let τ' be such that $\tilde{E}(\tau')\psi_1 - \tilde{E}g(\tilde{X}_{\tau'}) = \beta > 0$. Then $\psi^*\le \psi_2 = \tilde{E}g(\tilde{X}_{\tau'})/\tilde{E}(\tau') \le \psi_1 - \beta/\tilde{E}(\lambda) < \psi_1$.

Proof. $\tilde{E}(\tau')\psi_1 - \tilde{E}g(\tilde{X}_{\tau'}) = \beta$ implies $\psi_1 = [\tilde{E}g(\tilde{X}_{\tau'})+]/\tilde{E}(\tau')$ and hence that $\psi_1 - \beta/\tilde{E}(\tau') = \tilde{E}g(\tilde{X}_{\tau'})/\tilde{E}(\tau')$. But $\psi_1 - \beta/\tilde{E}(\tau') \le \psi_1 - \beta/\tilde{E}(\lambda)$. The result is established.

(3.5) *Lemma.* Let $\epsilon \le \alpha/2$. If $0 \le \psi_1 \le J$ and τ_ϵ are such that

$$\tilde{E}(\tau_\epsilon)\psi_1 - \tilde{E}g(\tilde{X}_{\tau_\epsilon}) \ge -\epsilon ,$$

then

$$\tilde{E}(\tau_\epsilon) \ge \alpha/2J .$$

Proof. $\tilde{E}(\tau_\epsilon)\psi_1 - \tilde{E}g(\tilde{X}_{\tau_\epsilon}) \geq -\epsilon$ implies that $\tilde{E}(\tau_\epsilon)\psi_1 \geq \tilde{E}g(\tilde{X}_{\tau_\epsilon}) - \epsilon \geq \alpha/2$. Thus, $\tilde{E}(\tau_\epsilon) \geq \alpha/2\psi_1 \geq \alpha/2J$.

(3.6) *Lemma.* Suppose $0 < \psi_1 \leq \psi^*$ and there exists τ_ϵ such that

$$\tilde{E}(\tau_\epsilon)\psi_1 - \tilde{E}g(\tilde{X}_{\tau_\epsilon}) \geq -\epsilon \tag{3.01}$$

where $\epsilon \leq \alpha/2$. Then $\psi_1 - \psi^* \geq -2\epsilon J/\alpha$.

Proof. From (3.01) we have that

$$\tilde{E}(\tau_\epsilon)(\psi_1-\psi^*) \geq \tilde{E}(\tau_\epsilon)(\psi_1-\psi^*) + \psi^*\tilde{E}(\tau_\epsilon) - \tilde{E}g(\tilde{X}_{\tau_\epsilon}) \geq -\epsilon \quad .$$

Thus

$$\psi_1 - \psi^* \geq -\epsilon/\tilde{E}(\tau_\epsilon) \geq -2\epsilon J/\alpha \quad ,$$

where the last inequality follows from (3.5).

These next two theorems show that solving the generalized ϵ-optimal replacement problem is equivalent to finding a b such that the b-maximization problem has a solution close enough to zero.

(3.7) *Theorem.* Let $\epsilon \leq 1$. If $\left|\sup_{\tau\leq\lambda} [\tilde{E}(\tau)\psi_1-\tilde{E}g(\tilde{X}_\tau)]\right| \leq \epsilon\alpha/2(J+1)$, then $|\psi_1 - \psi^*| \leq \epsilon$.

Proof. This follows from (3.3) and (3.6).

(3.8) *Theorem.* If $|\psi_1-\psi^*| < \epsilon$ then $\left|\sup_{\tau\leq\lambda} [\tilde{E}(\tau)\psi_1-\tilde{E}g(\tilde{X}_\tau)]\right| \leq \epsilon\tilde{E}(\lambda)$.

Proof. We have $\tilde{E}(\tau)\psi_1 - \tilde{E}g(\tilde{X}_\tau) = \tilde{E}(\tau)(\psi_1-\psi^*)+\tilde{E}(\tau)\psi^*-\tilde{E}g(\tilde{X}_\tau)$. If $\psi_1-\psi^* \geq 0$ then

$$\tilde{E}(\tau)\psi^* - \tilde{E}g(\tilde{X}_\tau) \leq \tilde{E}(\tau)(\psi_1-\psi^*) \leq \tilde{E}(\lambda)\epsilon \ .$$

If $\psi_1-\psi^* < 0$, let τ_n be $1/n$-optimal for $\psi^*\tilde{E}(\tau)-\tilde{E}g(\tilde{X}_\tau)$. Then,

$$\tilde{E}(\tau_n)\psi_1 - \tilde{E}g(\tilde{X}_{\tau_n}) \geq \tilde{E}(\tau_n)(\psi_1-\psi^*) - 1/n$$

$$\geq \tilde{E}(\lambda)(\psi_1-\psi^*) - 1/n \ .$$

Since n is arbitrary, it follows that

$$\sup_{\tau\leq\lambda} [\tilde{E}(\tau)\psi_1 - \tilde{E}g(\tilde{X}_\tau)] \geq \tilde{E}(\lambda)(\psi_1-\psi^*) \geq -\epsilon\tilde{E}(\lambda) \ .$$

This next result should be viewed as a corollary to (3.3) rather than (3.7) above. Note that we do not require $\epsilon \leq 1$.

(3.9) *Corollary*. Suppose that $\psi_1 = \tilde{E}g(\tilde{X}_\tau)/\tilde{E}(\tau)$ for some $\tau\in T$ and that τ' is an $\epsilon\alpha/4(J+1)$-optimal time for the ψ_1-maximization problem such that $\psi_1\tilde{E}(\tau') - \tilde{E}g(\tilde{X}_{\tau'}) \leq \epsilon\alpha/4(J+1)$. Then τ is an ϵ-optimal policy.

Consider the following iterative scheme:

(3.10) *(Iterative Scheme)*.

Initialize $\psi = J$ and $\tau = \lambda$. Find an $\epsilon\alpha/4(J+1)$-optimal time τ' for the ψ-maximization problem. If $\psi\tilde{E}(\tau')-\tilde{E}g(\tilde{X}_{\tau'}) \leq \epsilon\alpha/4(J+1)$ then τ is ϵ-optimal and we stop. Otherwise, reset ψ to $\tilde{E}g(\tilde{X}_{\tau'})/\tilde{E}(\tau')$ and τ to τ' and repeat the procedure.

From (3.9) and (3.4) we see that this iterative scheme will terminate finitely and yield an ϵ-optimal policy.

We have thus reduced the problem of finding generalized approximately optimal replacement policies to the problem of finding, for each $b > 0$, approximately optimal stopping times for the b-maximization problem.

IV. APPLICATION TO CERTAIN MARKOVIAN MODELS

In view of the results of the preceeding section, one approach to solving an optimal replacement problem (X,λ,g) is to discretize the time parameter and the state space obtaining an approximating stochastic process X' and a new problem (X',λ',g') where the λ' and g' are obtained from λ and g in some natural manner. By further reducing to a process X'' with $0 \leq t \leq N$, where N is some large integer, and with only a finite number of states one can then apply backwards induction, c.f., [8] Chapter 3, to obtain an optimal solution to any given b-maximization problem associated with the finite problem (X'',λ'',g''), here again λ'' and g'' are to be obtained in some natural way from λ' and g', respectively. Provided these two reductions can be justified, the optimal solution to the (X'',λ'',g'') b-maximization problem will be approximately optimal for the (X,λ,g) b-maximization problem, thus providing a means for applying the iterative technique (3.10).

In order to illustrate the approach just outlined, we now apply a version of it to the case of certain Markovian one dimensional wear processes with threshold lifetimes. In this setting we are able to give a theoretical justification of the required approximations. The spirit of these arguments is similar to the work of Whitt in [20] although the problems we are considering do not fit into the framework of that paper. At the end of the section, we specialize further to an exponentially distributed compound Poisson model and actually implement the procedure.

Throughout the rest of this section, unless stated otherwise, $X = (\Omega, X_t, M_t, M, P^x, \phi_t)$ is a nonterminating, Hunt wear/damage process. That is, a Hunt process which is nondecreasing, non-negative, real valued and nonterminating. For the basic theory concerning Hunt processes, see [7], chapter 1. Assume that for each $\zeta \in (0,\infty)$, the Markov time $\overline{\zeta} = \inf\{t: X_t \geq \zeta\}$ satisfies:

(4.01) $E^x(\overline{\zeta})$ is finite for all $0 \leq x \leq \zeta$, is right continuous and has left-hand limits in x.

Let $\lambda = \overline{\zeta}$ be such a stopping time for X. We let B_o denote the class of bounded, real valued, functions g on E^Δ which are Borel measurable on $[0,\infty)$ and are right continuous with left-hand limits.

For each such g, the *optimal stopping problem at* $x(st(X,\lambda,g))$ is to maximize $\tilde{E}^x g(\tilde{X}_\tau)$ over all Markov times τ and, if possible, find an optimal time τ_0. The Markov time τ_ϵ is called an ϵ-optimal time if $\tilde{E}^x g(\tilde{X}_{\tau_\epsilon}) \geq \Pi_{\tilde{X}} g(x) - \epsilon$ where $\Pi_{\tilde{X}} g(x) = \sup_\tau E^x g(\tilde{X}_\tau)$. The problem $st(X,\lambda,g)$ will also be denoted by $st(\tilde{X},g)$. In the case of Markov chains like those of section 2, if $\lambda = \inf\{n | Y_n = \Delta\}$, then $\tilde{Y}$ can be identified with Y and we will use $st(Y,g)$ to denote the optimal stopping problem. Furthermore, $\Pi_Y g(x)$ will denote $\sup_\tau E^x g(Y_\tau)$.

The following result is proved using standard techniques of advanced calculus:

(4.1) *Lemma.* Let $g \in B_o$. Then for each $\alpha > 0$ there exists a partition $0 = x_0 < x_1 < \ldots < x_m = \zeta$ of the interval $[0,\zeta)$ into subintervals of the form $[x_{i-1}, x_i)$ such that $|g(x) - g(x_{i-1})| < \alpha$ for all $x \in [x_{i-1}, x_i)$.

For $g \in B_o$ define $||g||$ by $||g|| = \sup_{x \in E^\Delta} |g(x)|$. This next result has a straight forward proof:

(4.2) *Lemma*. If $g_1, g_2 \in B_o$ with $||g_1 - g_2|| < \alpha$, then any ϵ-optimal time for the problem $st(X,\lambda,g_1)$ is $\epsilon + \alpha$ optimal for the problem $st(X,\lambda,g_2)$.

Let $g \in B^o$. Using the well known fact (c.f., [10]) that, for $\tau \le \lambda$, $\tilde{E}^o(\tau) = \tilde{E}^o(\tilde{E}^o(\lambda) - \tilde{E}^{X_\tau}(\lambda))$, we see that the b-maximization probelm associated with the problem (X,λ,g) can be converted into the optimal stopping problem $st(X,\lambda,G)$ where

$$G(x) = b\tilde{E}^o(\lambda) - b\tilde{E}^x(\lambda) - g(x), \text{ for } x \in [0,\zeta) ,$$

and

$$G(\Delta) = b\tilde{E}^o(\lambda) - g(\Delta) .$$

Furthermore, if $g \in B^o \cap B_o$ then, under assumption (4.01), $G \in B_o$.

We now consider the problem of finding arbitrarily close to optimal times for problems $st(X,\lambda,g)$ where $g \in B_o$. We may consider the killed process $\tilde{X}$ to be a nonterminating, Hunt process by considering its state space to be the one point compactification of $[0,\zeta)$ with the point at infinity denoted by Δ. Denote by M' the class of all stopping times for $\tilde{X}$. It is clear that, for $\tau \in M'$, $\tilde{E}^x g(\tilde{X}_\tau) = \tilde{E}^x g(\tilde{X}_{\tau\wedge\lambda})$. Hence $\sup_{\tau\in M'} \tilde{E}^x g(\tilde{X}_\tau) = \sup_{\tau\le\lambda} \tilde{E}^x g(\tilde{X}_\tau)$.

For each $t_o > 0$, associate with the process $\tilde{X}$ the Markov chain $t_o(\tilde{X}) = t_o(X)$ defined on Ω be $t_o(X)_n = \tilde{X}_{nt_o}$. Clearly $t_o(X)$ has transition probability $P(x,A) = \tilde{P}^x(\tilde{X}_{t_o} \in A)$. For each $t_o > 0$ let $S(t_o) = \{0, t_o, 2t_o, \ldots\}$ and let τ be an $S(t_o)$ valued stopping time for $\tilde{X}$. Then $\underline{\tau} = \tau/t_o$ is a stopping time for $t_o(X)$ and this correspondence is one-to-one.

This next result follows from Lemma 3, p. 118 of [18].

(4.3) *Lemma.* Let $g \in B_o$ and let $\epsilon > 0$. There exists a $t_o = 2^{-n_o}$ such that $\Pi_{t_o(X)} g(x) \geq \Pi_{\tilde{X}} g(x) - \epsilon/2$.

It is obvious that if τ is an ϵ-optimal $S(t_o)$-valued time for the problem $st(\tilde{X},g)$ then $\underline{\tau}$ is an ϵ-optimal time for the problem $st(t_o(X),g)$.

(4.4) *Lemma.* Let $g \in B_o$ and $\epsilon > 0$. For each x there exists an $M > 0$ such that, if $\tau \leq \lambda$ is $\epsilon/2$-optimal for $st(\tilde{X},g)$ then $\tau \wedge M$ is ϵ-optimal for $st(\tilde{X},g)$.

Proof. Let M be such that $\tilde{P}^x(\lambda > M) < \epsilon/(4(||g||+1))$. Then

$$\tilde{E}^x g(\tilde{X}_{\tau\wedge M}) = \tilde{E}^x g(\tilde{X}_\tau) I_{\{\tau \leq M\}} + \tilde{E}^x g(\tilde{X}_M) I_{\{\tau > M\}}$$

$$\geq \tilde{E}^x g(X_\tau) I_{\{\tau \leq M\}} - ||g|| \tilde{P}^x(\lambda > M)$$

$$\geq \tilde{E}^x g(X_\tau) I_{\{\tau \leq M\}} + \tilde{E}^x g(\tilde{X}_\tau) I_{\{\tau > M\}} - \epsilon/2$$

$$= \tilde{E}^x g(\tilde{X}_\tau) - \epsilon/2 \geq \Pi_{\tilde{X}} g(x) - \epsilon .$$

The result is proved.

The problem of finding $\sup_{\tau \leq N} \tilde{E}^x g(\tilde{X}_\tau) \equiv \Pi^N_{\tilde{X}} g(x)$ will be referred to as the N-bounded optimal stopping problem and denoted by $st(\tilde{X},g,N)$. A time $\tau_\epsilon \leq N$ is ϵ-optimal for $st(\tilde{X},g,N)$ if $\tilde{E}^x g(\tilde{X}_{\tau_\epsilon}) \geq \Pi^N_{\tilde{X}} g(x) - \epsilon$.

(4.5) *Lemma.* Let $g \in B_o$ and $\epsilon > 0$. For each x there exists a $t_o = 2^{-n_o}$ and $M > 0$ such that, for $N = M/t_o$

$$\Pi^N_{t_o(X)}g(x) \geq \Pi_{t_o(X)}g(x) - \epsilon/4 \geq \Pi_{\tilde{X}}g(x) - 3\epsilon/4 .$$

Proof. Let $t_1 = 2^{-n_1}$ be chosen so that $\Pi_{t_1(X)}g(x) \geq \Pi_{\tilde{X}}g(x) - \epsilon/2$. Now, using $\epsilon/8$ in (4.3), choose $t_o = 2^{-n_o} \leq t_1$, such that $\Pi_{t_o(X)}g(x) \geq \Pi_{\tilde{X}}g(x) - \epsilon/16$. Let τ be $\epsilon/16$-optimal for $st(t_o(X),g)$, then $\bar{\tau} = t_o\tau$ is $\epsilon/8$-optimal for $st(\tilde{X},g)$. Choose, by virtue of (4.4), a M such that $\bar{\tau}\wedge M$ is $\epsilon/4$-optimal for $st(\tilde{X},g)$. Since $\underline{\bar{\tau}\wedge M} = \tau\wedge(M/t_o)$ must be $\epsilon/4$-optimal for $st(t_o(X),g)$, we must have that, for $N = M/t_o$,

$$\Pi^N_{t_o(X)}g(x) \geq E^x_{t_o(X)}g(t_o(X)_{\tau\wedge N}) \geq \Pi_{t_o(X)}g(x) - \epsilon/4$$

$$\geq \Pi_{t_1(X)}g(x) - \epsilon/4 \geq \Pi_{\tilde{X}}g(x) - 3\epsilon/4.$$

The result is proved.

(4.6) *Theorem*. Let $g\in B_o$ and $\epsilon > 0$. For each x there exist a $\beta > 0$ and an $M > 0$ such that, if $0 < t_o \leq \beta$, then any $\epsilon/8$-optimal time for the $N = M/t_o$ bounded problem $st(t_o(X),g,N)$ will be ϵ-optimal for the problem $st(\tilde{X},g)$.

Proof. Using (4.5), let $t_1 = 2^{-n_1}$ and $N_1 = M_1 2^{n_1}$ be chosen so that

$$\Pi^{N_1}_{t_1(X)}g(x) \geq \Pi_{t_1(X)}g(x) - \epsilon/4 \geq \Pi_{\tilde{X}}g(x) - 3\epsilon/4 .$$

Let τ_o be 0-optimal for $st(t_1(X),g,N_1)$. For each $k = 0,1,\ldots,[N_1]$, let $\alpha_k > 0$ and $A_k\subset\tilde{\Omega}$ be such that,

$$\tilde{P}^x(A_k) < \epsilon/[32(||g||+1)([N_1] + 1]$$

and

$$|g(\tilde{X}_r) - g(\tilde{X}_{k/2^{n_1}})| \le \epsilon/16$$

for all $\omega \notin A_k$ and for $r \in [k/2^{n_1}, k/2^{n_1} + \alpha_k)$.

Let $\beta = \min\{\alpha_k, 2^{-n_1}\}$ and let $0 < t_o < \beta$. Define τ piecewise on $\{\tau_o = k\}$ by $\tau(\omega) = mt_o$ where $m = \min\{n | nt_o \ge k/2^{n_1}\}$. Clearly, $\tau \in S(t_o)$ and $\tau \le M_1 + 2^{-n_1} < M_1 + 1$. Let $M = M_1 + 1$, then $\underline{\tau} \le N = M/t_o$ and

$$\Pi^N_{t_o(X)} g(x) \ge E^x_{t_o(X)} g(t_o(X)_{\underline{\tau}}) = \tilde{E}^x g(\tilde{X}_\tau) \ge \tilde{E}^x g(\tilde{X}_{\overline{\tau}_o}) - \epsilon/8 .$$

The last inequality follows by splitting $g(\tilde{X}_{\tau_o})$ up on $\Omega - \bigcup_k A_k$ and $\bigcup_k A_k$. But,

$$\tilde{E}^x g(\tilde{X}_{\tau_o}) - \epsilon/8 = E^x_{t_1(X)} g(t_1(X)_{\tau_o}) - \epsilon/8$$

$$= \Pi^{N_1}_{t_1(X)} g(x) - \epsilon/8$$

$$\ge \Pi_{\tilde{X}} g(x) - 7\epsilon/8 .$$

Thus,

$$\Pi^N_{t_o(X)} g(x) \ge \Pi_{\tilde{X}} g(x) - 7\epsilon/8$$

and the result follows from this.

To this point, we have reduced the problem of finding approximately optimal times for $st(\tilde{X},g)$ to finding approximately optimal times for $st(t_o(X),g,N)$ for small enough t_o and large enough N. The form of the optimal solution to such problems is well known, c.f. [18], chapter 2. In order to approximate the solution in the general case, however, it is necessary to discretize the state space, and to justify this last reduction, we make three more assumptions on the process X:

(i) For each $t_o > 0$, $P^x(X_{t_o} = x)$ is continuous in x.

(ii) For $t_o > 0$, and $\epsilon > 0$ there exists a $\delta > 0$ such that $P^x(X_{t_o} \in (x,x+\delta)) < \epsilon$.

(iii) For $t_o > 0$ and $\epsilon > 0$ there exists $\delta > 0$ such that, if $|x-y| < \delta$ and $A \subset [0,\infty)\setminus\{x,y\}$ then $|P^x(X_{t_o} \in A) - P^y(X_{t_o} \in A)| < \epsilon$.

Now, consider the problem $st(\tilde{X},g)$ where $g \in B_o$ is a right-continuous step function. Let $\epsilon > 0$ and $t_o > 0$ be given and choose, by virtue of (2.3), a $\gamma > 0$ such that,

$$|E^x_Z g(t_o(X)_\tau) - E^x_{t_o(X)} g(t_o(X)_\tau)| \le \epsilon/24$$

whenever $d(Z,t_o(X)) \le \gamma$. Recall here that $Z \in S_{t_o}(X)$ and therefore $Z_n(\omega) = t_o(X)_n(\omega)$ for all $\omega \in \tilde{\Omega}$ and $n = 0,1,2,\ldots,$.

Let $\delta_1 > 0$ be chosen so that, if $|x-y| < \delta_1$, then $|P^x(X_{t_o} = x) - P^y(X_{t_o} = y)| < \gamma/5$. Let $\delta_2 > 0$ be such that

$P^x(X_{t_o} \in (x,x+\delta_2)) < \gamma/5$ for all x. Finally, let $\delta_3 > 0$ be chosen so that if $|x_1-x_2|<\delta_3$ and $A\subset E\backslash\{x_1,x_2\}$ then $|P^{x_1}(X_{t_o} \in A) - P^{x_2}(X_{t_o} \in A)| < \gamma/5$.

Choose a partition $0 = x_0 < x_1 <\ldots<x_m = \xi$ of $[0,\zeta)$ so that each interval has length less than or equal to $\min\{\delta_1,\delta_2,\delta_3\}$ and each of the jump points of g is a partition point. Consider the transition probability $\hat{P}(x,\cdot)$ on $[0,\zeta)\cup\{\Delta\}$ which is determined by the following: for $x\in[x_{i-1},x_i)$,

$$\hat{P}(x,A) = P(x_{i-1},[x_{i-1},x_i)) + P(x_{i-1},A\cap[x_i,\zeta)), \quad \text{if} \quad x\in A$$

$$\text{and} \quad A\subset[0,\zeta) .$$

$$\hat{P}(x,A) = P(x_{i-1},A\cap[x_i,\zeta)) \quad \text{if} \quad x\notin A,\ A\subset[0,\zeta),\ \hat{P}(x,\Delta)$$

$$= P(x_{i-1},\Delta).$$

If $x = \Delta$ then $\hat{P}(\Delta,\Delta) = 1$.
For $\hat{P}$ we have

$$\begin{aligned}|\hat{P}(x,A) - P(x,A)| \le{}& |P(x_{i-1},x_{i-1}) - P(x,x)| \\ &+ P(x_{i-1},(x_{i-1},x_i)) + P(x,(x,x_i)) \\ &+ |P(x_{i-1},A\cap[x_i,\zeta)) - P(x,A\cap[x_i,\zeta))| \\ &+ |P(x_{i-1},\Delta) - P(x,\Delta)| \le 5\gamma/5 = \gamma.\end{aligned}$$

To see this, consider first the case $x \in A$. Then

$$\hat{P}(x,A) = P(x_{i-1},x_{i-1}) + P(x_{i-1},(x_{i-1},x_i))$$

$$+ P(x_{i-1},A\cap[x_i,\zeta) + P(x_{i-1},A\cap\{\Delta\})$$

and

$$P(x,A) = P(x,x) + P(x,A\cap(x,x_i))$$

$$+ P(x,A\cap[x_i,\zeta)) + P(x,A\cap\{\Delta\}) \quad .$$

Thus $|\hat{P}(x,A) - P(x,A)| \leq |P(x_{i-1},x_{i-1}) - P(x,x)|$

$$+ P(x_{i-1},(x_{i-1},x_i)) + P(x,(x,x_i)\cap A)$$

$$+ |P(x_{i-1},A\cap[x_i,\zeta)) - P(x,A\cap[x_i,\zeta))|$$

$$+ |P(x_{i-1},A\cap\{\Delta\}) - P(x,A\cap\{\Delta\})| \ .$$

Since the last term either equals 0 or

$|P(x_{i-1},\Delta) - P(x,\Delta)|$ and since $P(x,A\cap(x,x_i)) \leq P(x,(x,x_i))$ we see that the desired inequality holds.

A similar argument establishes that, in the case $x \notin A$,

$$|\hat{P}(x,A) - P(x,A)| \leq P(x,(x,x_i))$$

$$+ |P(x_{i-1},A\cap[x_i,\zeta)) - P(x,A\cap[x_i,\zeta))|$$

$$+ |P(x_{i-1},\Delta) - P(x,\Delta)| \quad .$$

Let Z be the Markov chain on Ω with the transition probability $\hat{P}$ and the same initial distributions as $\tilde{X}$.

(4.7) *Lemma*. Let t_o, Z be as above and let N be given. Then any $\epsilon/24$-optimal time for $st(Z,g,N)$ is $\epsilon/8$-optimal for $st(t_o(X),g,N)$.

Proof. First, let τ_o be optimal for $st(t_o(X),g,N)$. Then

$$\Pi_Z^N g(x) \geq E_Z^x g(t_o(X)_{\tau_o}) \geq E_{t_o(X)}^x g(t_o(X)_{\tau_o}) - \epsilon/24$$

$$= \Pi_{t_o(X)}^N g(x) - \epsilon/24 .$$

Now, let τ' be an $\epsilon/24$-optimal time for $st(Z,g,N)$. Then

$$\begin{aligned} E_{t_o(X)}^x g(t_o(X)_{\tau'}) &\geq E_Z^x g(t_o(X)_{\tau'}) - \epsilon/24 \\ &\geq \Pi_Z^N g(x) - \epsilon/12 \\ &\geq \Pi_{t_o(X)}^N g(x) - \epsilon/8 . \end{aligned}$$

The lemma is proved.

Let W be the discrete Markov chain with finite state space $S = \{0,x_1,x_2,\ldots,x_{m-1},\Delta\}$ and transition probability $\overline{P}(x_i,x_j) = \hat{P}(x_i,[x_j,x_{j+1}))$, $\overline{P}(x_i,\Delta) = \hat{P}(x_i,\Delta)$ and $\overline{P}(\Delta,\Delta) = 1$. Define $\overline{g}$ on S be $\overline{g}(y) = g(y)$.

For $x \in [0,\zeta)$ define $e(x) = x_i$ if $x \in [x_i,x_{i+1})$. Take $e(\Delta) = \Delta$. If h is a function which is constant on each of the intervals $I_j = [x_{j-1},x_j)$, then note that

$$\hat{E}^x h(t_o(X)_1) = \overline{E^x \bar{h}}(W_1) .$$

It follows that $\hat{E}^x h(t_o(X)_1)$ is constant on the intervals I_j as well. But then so is $\hat{E}^x h(t_o(X)_1) \vee h(x) = \overline{E}^{e(x)} \bar{h}(W_1) \vee \bar{h}(e(x))$.

For a Markov chain Y and function g on the state space of Y define the operators $Q^n g(x)$ by

$$Q^o g(x) = g(x)$$

$$Q^{n+1} g(x) = E_Y^x Q^n g(Y_1) \vee Q^n g(x) .$$

From the preceding discussion, we have that $\hat{Q}^n h(x) = \overline{Q^n \bar{h}}(e(x))$ for all $n = 0,1,2,\ldots,$. Given N, the optimal solutions to the problems $st(Z,g,N)$ and $st(W,\bar{g},N)$ are given by Theorem 4, p. 28 of [18]. From the above relations between $\hat{Q}^n$ and $\overline{Q}^n$, we see that the solution to the problem $st(Z,g,N)$ can be obtained from that for $st(W,\bar{g},N)$ as follows: Let

$$C_m = \{x \in S: \overline{Q^{N-m}} \bar{g}(x) = \bar{g}(x)\}$$

and let $\hat{C}_m = e^{-1}(C_m)$. The optimal stopping time for $st(Z,g,N)$ is to stop as soon as $t_o(X)_m \in \hat{C}_m$. If N and t_o have been chosen according to (4.7), then this time will be at least $\epsilon/24$-optimal for $st(Z,g,N)$, hence, $\epsilon/8$-optimal for $st(t_o(X),g,N)$ and thus ϵ-optimal for $st(\tilde{X},g)$.

(4.8) *Remark*. The results of this section have fairly straightforward generalizations to the setting of Markovian, multivariate wear processes. In particular, this allows the incorporation of the time dimension into the state space which enables these techniques to be used for nonhomogeneous problems. More importantly, the cost functions may be time dependent which considerable expands the scope of optimal replacement applications.

(4.9) *Remark*. Beside the errors due to discretization, it remains to handle errors due to imperfect knowledge about the original model. This is addressed in [17].

We implemented the procedure discussed in the last two sections for compound Poisson processes with rate perameter λ and with damage imcrements distributed according to an exponential distributions with parameter μ. The program was then run for the case $\mu = \lambda = 1$, threshold $\zeta = 5$ and cost functions g which took value 1 on $[0,5)$ and 2, 11 and 31, respectively, at Δ. Two of these cases were analyzed by Taylor in [19] where it was shown that the long run average is .27 for $g(\Delta) = 2$ and .5 for $g(\Delta) = 11$.

For these runs, the program chopped the time interval [0, 15.94] into 400 equal subintervals and chopped [0,5) into 40 equal subintervals. The results of these runs are given below. They tend to indicate how fast the interative scheme (3.10) terminates. The data also indicates the dependence of the accuracy of approximation, for a given mesh, on the magnitude of the difference between the cost at failure and the cost of a planned replacement.

A = Average time to repair NI = Number of Iterations

$g(\Delta)$	ψ^*	A	ψ_0	ψ_1	$E^o(\tau_1)$	ψ_2	$E^o(\tau_2)$
2	.27	4.69	.67	.27	4.96	.27	4.70
11	.50	3.00	1.83	.67	4.32	.52	3.30
31	.76	2.32	5.17	1.52	4.32	.88	3.04

$g(\Delta)$	ψ^*	ψ_3	$E^o(\tau_3)$	ψ_4	$E^o(\tau_4)$	NI
2	.27	----	----	---	----	2
11	.50	.51	3.04	---	----	3
31	.76	.80	2.40	.80	2.28	4

When $g(\Delta) = 31$, one gets within .01 of the true value by dividing [0,15.94] into 1600 intervals and [0,5) into 100 intervals.

REFERENCES

1. Abdel-Hameed, M.S., Optimal replacement policies for devices subject to a gamma wear process, "The Theory and Applications of Reliability with Emphasis on Bayesian and Non-Parametric Methods", Vol. 1, ed. by C.P. Tsokos and I.N. Shimi, Academic Press, 397-412, (1977).
2. Abdel-Hameed, M.S., Life distribution properties of devices subject to a Levy wear process, to appear, *Math. Oper. Res.*
3. Abdel-Hameed, M.S. and Proschan, F., Nonstationary Shock Models, *Stoch. Proc. Appl.*, 383-404, (1973).
4. Abdel-Hameed, M.S. and Shimi, I.N., Optimal replacement of damaged devices, *J. Appl. Prob.*, *15*, 153-161 (1978).
5. Barlow, R., and Proschan, F., "Mathematical Theory of Reliability", John Wiley & Sons, New York, (1965).
6. Bergman, B., Optimal Replacement under a General Failure Model, *Adv. Appl. Prob.*, *10*, 431-451, (1978).
7. Blumenthal, R.M. and Getoor, R.K., "Markov Processes and Potential Theory", Academic Press, (1969).
8. Chow, Y.S., Robbins, H. and Siegmund, D., "Great Expectation: The Theory of Optimal Stopping", Houghton Mifflin, Boston, (1971).
9. Cox, D.R., "Renewal Theory", Methuen, London, (1962).
10. Dynkin, E.B., The optimum choice of the instant for stopping a Markov Process, Doklady Academii Nauk, *SSSR*, 150(2), 238-240, (1963).
11. Feldman, R.M., Optimal replacement with Semi-Markov Shock Models, *J. Appl. Prob.*, *13*, 108-117, (1976).
12. ----------, Optimal replacement with Semi-Markov Shock Models using discounted costs, *Math. Oper. Res.*, (1976).

13. ----------, Optimal replacement of systems governed by Markov additive processes, *Annals of Prob.*, *5*, 413-429 (1972).

14. Fontenot, R., and Proschan, F., *Imperfect Maintenance*, in preparation.

15. Gottlieb, G., *Optimal Replacement for Shock Models with Generalized Failure Rate*, preprint.

16. Gottlieb, G., and Levikson, B., *Optimal Replacement for Self-Repairing Shock Models with General Failure Rate*, preprint.

17. Quinn, J., *Stability of Optimal Replacement Policies*, preprint.

18. Shiryayev, A. N., "Optimal Stopping Rules, Applications of Mathematics, E," Springer-Verlag, New York, Heidelberg, Berlin (1978).

19. Taylor, H. M., Optimal replacement under additive damage and other failure models, *Naval Res. Log. Quart.*, *22*, 1-18 (1975).

20. Whitt, W., Approximations of dynamic programs, I. *Math. Oper. Res.*, *3*, 231-243 (1978).

21. Zuckerman, D., Optimal stopping in a semi-Markov shock model, *J. Appl. Prob.*, *15*, 629-634 (1978).

22. Zuckerman, D., Optimal replacement policy for the case where the damage is a one-sided Lévy process, *Stoch. Proc. & Appl.*, *7*, 141-151 (1978).

OPTIMAL INVENTORIES FOR OVERHAUL OF REPAIRABLE REDUNDANT SYSTEMS: A MARKOV DECISION MODEL

Margaret K. Schaefer

Department of Mathematics and Computer Science
College of William and Mary
Williamsburg, Virginia

Many complex machines which must perform very reliably are designed with a k-out of -n component configuration, in which the system is operable so long as at least k of its n identical components work. Such machines are often composed of repairable components and are maintained through scheduled, periodic overhauls, at which all failed parts are replaced by good ones. We develop a Markov decision model to determine the optimal inventories of repairable spare parts for such a system and simultaneously determine the optimal repair policy when there is a choice at each overhaul among a set of repair rates. The repair policy may be state-dependent; for instance, a policy might be to use a fast repair rate when q or fewer spares are on hand at the end of an overhaul, and use a slow repair rate otherwise. The objective is to minimize total expected costs of spares, stockouts and repairs. A simple example is included to illustrate the model and its solution through linear programming.

[1]*This research was supported by grant NSG 1625 from the NASA-Langley Research Center.*

ISBN 0-12-041420-1

I. INTRODUCTION

Many types of complex machines which must perform very reliably are designed so that they can function even when some components have failed, employing a k-out-of-n component configuration. Often such machines are scheduled for periodic overhauls at a maintenance center, during which all failed parts are replaced by good ones so that the machine's reliability is renewed. In general, periodic overhauls are called for when emergency repair during the time interval between overhauls is impossible or very costly; this will often be the case for systems on submarines, helicopters, and space vehicles.

A typical application of the model involves ultra-reliable avionics control systems for commercial aircraft, presently being developed by NASA. The avionics system will consist of self-contained modules, each having several computer components, which are reliable and expensive. Inventories of these repairable modules will be stocked at a maintenance center to meet demands at periodic overhauls.

To meet the extremely high reliability requirements (e.g., a maximum failure rate of 10^{-10} per hour avionics application), the machines leave the maintenance center in "good-as-new" condition. Consequently, if a needed part or parts are not available in current inventory, the overhaul cannot be completed without incurring extra cost. Required parts which are out of stock must be obtained by some emergency procedure, such as borrowing or priority repair of the part. Stockout costs will vary depending upon the number of parts required.

A model to determine the optimal repairable parts inventory and repair strategy for such a maintenance center is developed in this paper. Total expected shortage costs, repair costs, and holding costs are minimized for a machine containing a single system of redundant parts. The problem is formulated as a Markov decision process. Transition probabilities are calculated for each possible state and repair rate, and the optimal spare parts inventory and repair strategy is determined through linear programming.

II. RELATED MODELS IN THE LITERATURE

Although much has been written about fault-tolerant systems [1], [2], [7], emphasis has been placed upon reliability. There is very little literature on the problem of optimal maintenance center inventories. Schaefer [5] found optimal maintenance center inventories for complex machines under a job-completion criterion where a single stockout penalty was assessed if the overhaul could not be completed because of shortage. The model handled machines with different types of parts, but exact results were obtained only when part failure rates were low and there was only one part of each type. The model presented here assesses a different stockout penalty for each number of parts missing, summing the expected costs. Redundant systems are handled explicitly. Schaefer and Lawrence [6] found optimal inventories for machines containing a set of independent redundant systems but the repair rate for each type of part was constant, so that the decision variables were the initial number of spares of each type.

Our model uses a discrete Markov process to model on-hand inventory. Other examples of this technique may be found in Tainiter [8], who found the steady-state probabilities for an inventory of rental cars, where the time of rental is analogous to repair time, and in Hadley and Whitin [3], where it was advocated for any inventory system whose condition is reviewed only at discretely spaced intervals of time.

III. THE MODEL AND ITS LINEAR PROGRAMMING FORMULATION

Consider a maintenance center serving a set of identical machines containing a single k-out-of-n system of independent identical workloads. We seek the initial inventory of spare parts and repair strategy which minimizes expected stockout costs, repair costs and holding costs. We assume that each machine arrives for overhaul in functioning condition so that at least k parts are working and n-k or fewer have failed during the cycle. That is, the machine is so reliable that the probability that more than n-k parts have failed is negligible. This is a reasonable assumption since redundancy is employed to ensure high reliability. Each failed part is immediately sent to a repair shop, from which it eventually returns to replenish the stock of spares at the maintenance center. The total number of spares on hand and undergoing repair is a constant, s. Stockouts may occur if demand for spares at an overhaul exceeds the number of spares presently on hand. We assume that one machine arrives for overhaul each day, each machine started the cycle in good-as-new condition, and that even if the part is stocked out, the machine is released from overhaul without appreciable delay, although a shortage penalty is assessed. Thus there are no backorders.

We assume that after each overhaul the maintenance center can choose the repair rate to be employed by the repair shop until the next overhaul. The number of initial spares and the repair rate are in a sense substitutable; a small number of spares repaired rapidly may be as effective as a large number of spares repaired at a slow rate. In fact, if repairs were instantaneous, no spares at all would be required.

We assume that the faster the repair rate, the higher the repair shop cost. Otherwise, it would never be optimal to use a low repair rate. Although higher repair rates are more costly, if they are utilized, the expected repair time per spare is decreased so the chances of a stockout penalty at the next overhaul are diminished. The optimal repair strategy might be to choose a slow repair rate at all times when there are some spares on hand and to use a faster rate when present spares inventory is low, so as to decrease the chance of stockout at the next overhaul. Of course, the optimal number of spares and spares strategy will depend upon the relative costs of stockouts repairs and holding costs.

We introduce the following notation:

H = daily holding cost of a spare;

R = the number of possible exponential repair rates, $r=1,2,\ldots,R$;

K_r = cost per day of repairing at rate r;

λ = constant failure rate of each component, in hours;

T = cycle time between overhauls, in hours;

$a = 1 - \exp(-\lambda T)$ = probability that a single part has failed during cycle time T hours;

μ_r = exponential repair rate in days using repair rate r;

$b_r = 1 - \exp(-\mu_r)$ = probability that a part which was at the repair shop at the end of one day returns to the center by the end of the next day using repair rate r;

L_i = penalty cost for shortage of i spares (emergency repair or borrowing cost);

$Z_r(t,u)$ = conditional probability that t spares return to the maintenance center from the repair shop on a day that begins with u in the repair shop, using repair rate r, $0 \le t \le u \le s$;

$P_s(r)$ = matrix of transition probabilities with elements P_{ij}, representing the probability that j spares are on hand at the end of a day, given that i spares were on hand at the end of the previous day, the repair rate was r and there were s spares on hand initially.

The probability v_m that m parts are demanded at an overhaul, $0 \le m \le n-k$, is $v_m = \binom{n}{m} a^m (1-a)^{n-m}$. Also, $Z_r(t,u) = \binom{u}{t} b_r{}^t (1-b_r)^{u-t}$. Then $P_{ij}(r)$ for a given value of s may be calculated as

$$P_{ij}(r) = \sum_{m=\max\{0,i-j\}}^{\min \Sigma\{s-j,n-k\}} v_m Z_r(m+j-i,s-i) \text{ for } i = 0,1,\ldots,s$$

$$\text{and } j = k-n,\ldots,s.$$

Since there are no back orders, $P_{ij}(r) = P_{0j}(r)$ for $i=k-n,\ldots,-1$ and all j.
A negative number on hand, $i<0$, indicates a shortage of i parts.

Then the problem may be formulated as a linear programming problem [4] with decision variables y_{ir} where y_{ir} = P state ={i and reapir rate = r}. For a given value of s, we formulate the problem of minimizing long run expected average cost of repairs and shortages per unit of time as

$$\min E[C(s)] = \sum_{i=k-n}^{s} \sum_{r=1}^{R} K_r y_{ir} + \sum_{i=k-n}^{-1} \sum_{r=1}^{R} L_i y_{ir}$$

subject to

$$(1) \quad \sum_{i=k-n}^{s} \sum_{r=1}^{R} y_{ir} = 1,$$

$$(2) \quad \sum_{r=1}^{R} y_{jr} - \sum_{i=k-n}^{s} \sum_{r=1}^{R} y_{ir} P_{ij}(r) = 0 \text{ for } j = k-n,\ldots,s-1,$$

(3) $y_{ir} \geq 0$, $i=k-n,\ldots,s$; $r = 1,2,\ldots R$.

This problem is of reasonable size, having only $R(s+n-k+1)$ variables and $s+n-k+1$ constraints, besides the non-negativity constraints. There are $s+n-k+1$ possible states of the system and there are exactly that number of nonzero y_{ir} values in the optimal solution. If $y_{ir} > 0$, r is the repair rate to be used when there are i on hand. There is only one positive y_{ir} for each i so the optimal repair policy is deterministic [4].

The problem can be solved by the simplex method. Alternatively, it can be solved by a policy-improvement algorithm, which is a good choice for this problem since many possible policies can be rejected as clearly non-optimal, reducing the potential number of policies to be examined. In particular, no strategy with fast repair for $i=q$ and slow repair for $i=q-1$ would ever be optimal, since fast repair costs more than slow repair and there is greater chance of stockout at next overhaul in state $i=q-1$ than for state $i=q$.

A separate linear programming problem must be solved for each value of s, $s=1,2,\ldots$; the total expected repair and stock-out cost $E[C(s)]$ is added to the holding cost Hs to get total expected costs. The optimal value of s and its associated strategy is the one that minimizes total expected costs.

IV. ILLUSTRATIVE EXAMPLE

Consider a machine containing a 4-out-of-6 redundant system. We assume that the machine always arrives for overhaul functioning, so that no more than 2 components have failed.[1] The other ing, so that no more than 2 components have failed.[1] The other parameters are as follows:

$$a = 0.05,\ R = 2,\ b_1 = 0.2,\ b_2 = 0.6,\ L_{-1} = \$500,\ L_{-2} = \$800,$$

$$K_1 = \$50,\ K_2 = \$75.$$

The results are summarized in Table 1, for $s \le 4$. For each positive value of s, the optimal strategy is to use the fast repair rate only when there are no spares on hand.

As noted earlier, holding costs have not yet been considered. The daily holding costs, Hs, must be added to E[C(s)] to get total costs, TC(s). The range of H values for which each s value is optimal may be calculated easily by finding the breakeven points where E[C(s)] + Hs=E[C(s+1)] + H(s+1). The resulting intervals are shown in Table 2.

TABLE 1

Linear Programming Solutions

s	$y_{ik}*$	Optimal Strategy	E[C(s)]
0	$v_{-2} = .0305$ $v_{-1} = .2321$ $v_0 = .7351$	Emergency repair only, no spares on hand. Probabilities of stockout are merely the v_m measures.	\$140.45
1	$y_{-22} = .0064$ $y_{-12} = .0608$ $y_{02} = .3075$ $y_{11} = .6254$	Fast repair for i = -2,-1,0, slow repair for $i \ge 1$.	\$94.84

[1] *This assumption should be verified by calculating the probability that the number failed exceeds n-k. In our example, this probability is only 0.002.*

TABLE 1 CONT.

s	y_{ik*}	Optimal Strategy	E[C(s)]
2	y_{-22} = .0017	Fast repair for i = -2,-1,0, slow repair for i≥1.	$65.33
	y_{-12} = .0201		
	y_{02} = .1346		
	y_{11} = .4422		
	y_{21} = .4015		
3	y_{-22} = .0005	Fast repair for i = -2,-1,0, slow repiar for i≥1.	$55.48
	y_{-12} = .0070		
	y_{02} = .0557		
	y_{11} = .2340		
	y_{21} = .4003		
	y_{31} = .3026		
4	y_{-22} = .0002	Fast repair for i = -2,-1,0, slow repair for i≥1.	$51.95
	y_{-12} = .0024		
	y_{02} = .0212		
	y_{11} = .1047		
	y_{21} = .2597		
	y_{31} = .3679		
	y_{41} = .2440		

TABLE 2

H and TC(s) Intervals

s	H Values for Which s is Optimal	Resulting TC(s)
0	H > $45.61	TC(0) ≥ $140.45
1	$29.51 < H < $45.61	124.32 ≤ TC(1) < $140.45
2	$9.85 < H < $29.51	75.18 ≤ TC(2) < $124.32
3	$3.53 < H < $9.85	59.01 ≤ TC(3) < $75.18
4	H < $3.53	TC(4) ≤ $59.01

Thus, by considering holding costs last, a simple sensitivity analysis for H is apparent. If holding costs are difficult to estimate exactly, it may be easier to decide in which interval the H value lies instead. Likewise, if H increases or decreases, Table 2 indicates whether a change in s is required.

CONCLUSION

We have applied Markov decision process techniques to a new problem, simultaneously finding the optimal initial number of repairable spares for a maintenance center and the optimal repair strategy. The situation is analogous to the problem of determining simultaneously the number of servers and their service rates in a retail facility to minimize total cost of servers and expected lost sales through balking. The required assumptions are not very restrictive, and few parameters must be estimated. The model requires solution of a reasonable number of linear programming problems by the simplex method or a policy-improvement algorithm.

REFERENCES

1. Barlow, R. E., and Proschan, F., *Statistical Theory of Reliability and Life Testing: Probability Models*, Holt, Rinehart and Winston, New York, (1975).
2. Bellman, R., and Dreyfus, S., "Dynamic Programming and the Reliability of Multi-component Devices", *Operations Research*, Vol.6, No. 2 (Mar.-Apr. 1958), pp. 200-206.
3. Hadley, G., and Whitin, T., *Analysis of Inventory Systems*, Prentice-Hall, Englewood Cliffs, N.J., (1963).

4. Hillier, F., and Lieberman, G., *Introduction to Operations Research*, Holden-Day, San Francisco, (1980).

5. Schaefer, M. K., "A Multi-Item Maintenance Center Inventory Model for Low-Demand Repairable Items", *Management Science*, Vol. 29, No. 3, pp. 1062-1068, Sept. (1983).

6. Schaefer, M. K., and Lawrence, S. H., "Optimal Maintenance Center Inventories for Fault-Tolerant Repairable Systems", to appear in *Journal of Operations Management*.

7. Shooman, M. L., *Probabilistic Reliability: An Engineering Approach*, McGraw-Hill, New York, N.Y. (1968).

8. Tainiter, M., "Some Stochastic Inventory Models for Rental Situations", *Management Science*, Vol. II, No. 2, pp. 316-326, November, (1964).

Life Testing

NONPARAMETRIC ESTIMATION FROM ACCELERATED LIFE TESTS WITH RANDOM CENSORSHIP[1]

D.T. McNichols

Department of Statistics
Virginia Polytechnic Inst. and State University
Blacksburg, Virginia

W.J. Padgett

Department of Mathematics and Statistics
University of South Carolina
Columbia, South Carolina

For an accelerated life test, let $V_1, \ldots, V_k$ *denote* k *fixed accelerated stresses and let* V_0 *be the normal stress. It is assumed that the probability distributions corresponding to the accelerated stresses differ from the nonaccelerated life distribution only by a scale factor. A simple nonparametric consistent estimator of the life distribution at the normal stress is developed for randomly right-censored data. The estimator also applies to accelerated life test data for items with two*

[1]*This research was partially supported by the U.S. Air Force Office of Scientific Research under Grant No. AFOSR 81-0166.*

ISBN 0-12-041420-1

independent failure modes (or competing risks) at each stress level. An example is given to illustrate the estimation procedure.

1. INTRODUCTION

Often in reliability studies, accelerated testing is used when one wishes to estimate the life distribution of a device which has a rather long lifetime under normal use. Items are subjected to greater stress than that of the use condition, and from the observed failure data an estimate of the distribution function of the lifetime of the item under the use condition is formed.

Several nonparametric procedures have been proposed for estimation from accelerated life test data. Shaked, Zimmer, and Ball (1979) assumed a functional relationship between the distribution functions corresponding to any two different stress levels and showed that their model is a generalization (in a proper sense) of some parametric models. Steck, Zimmer, and Williams (1974), Shaked and Singpurwalla (1982), and Basu and Ebrahimi (1982) assumed a functional relationship between the accelerated and non-accelerated distribution functions, while Barlow and Scheuer's (1971) nonparametric model assumed a stochastic ordering between a single accelerated distribution and the nonaccelerated distribution.

With the exception of the results of Barlow and Scheuer (1971) and Basu and Ebrahimi (1982), the models discussed above were for complete samples. In many situations the observations may be randomly censored from the right. This occurs often in life testing where items may be removed from the study at various times for more extensive analysis or for other reasons. Another situation which may be treated in the same manner as randomly right-censored data is when there are two independent causes of failure, and the life distribution or other parameters for one of the failure modes is to be estimated. In this paper, a simple nonparametric estimator of the distribution function (and mean life) of an item in the nonaccelerated environment is developed based on randomly right-censored observations (or observed lifetimes under two independent causes of failure) from accelerated life tests. The estimator assumes that data are available from each of the nonaccelerated and $k \geq 1$ accelerated environments and is shown to be strongly consistent for the true life distribution. This paper extends the results of Barlow and Scheuer (1971), where randomly right-censored data were assumed to be available from the nonaccelerated environment in addition to a single accelerated life test. Also, Basu and Ebrahimi (1982) required the same acceleration model as that of Shaked, Zimmer and Ball (1979) whereas we do not.

Throughout, let F_V denote the life distribution of items on test under stress level V. Let $V_1,\ldots,V_k$ be stress levels under accelerated testing and let V_0 denote the normal use stress. Assume that k and the V_i's are selected at the outset and fixed. Let X_i^o denote the true lifetime random variable at stress V_i. It is assumed that under stress V_i there is a censoring variable U_i having distribution H_i which is independent of X_i^o and is independent of other life tests at different stress levels, that is, $U_0,U_1,\ldots,U_k$ are independent. Furthermore, it is assumed that for increased stress, F_{V_i} and H_i differ from F_{V_0} and H_0, respectively, only by a scale change for $i=1,\ldots,k$. That is, there exist positive constants θ_i, $i=1,\ldots,k$ such that $\theta_i X_i^o$ and $\theta_i U_i$ have the same distributions as X_0^o and U_0, respectively. In the case of two independent causes of failure, X_i^o denotes the lifetime of the item at stress V_i due to the "main" cause of failure (or the failure mode of interest) and U_i represents the lifetime at V_i due to the other cause. In either situation the distribution function F_{V_0} of X_0^o is to be estimated.

Let X^o_{ij} and U_{ij} denote the true failure time and the "censoring" time, respectively, $j=1,\ldots,n_i$, for each of the n_i items on test under stress level V_i, $i=0,1,\ldots,k$. The observations obtained by the experimenter at stress level V_i are (X_{ij},Δ_{ij}), $j=1,\ldots,n_i$, where

$$X_{ij} = \min(X^o_{ij}, U_{ij}),\ \Delta_{ij} = \begin{cases} 1 & \text{if } X^o_{ij} \le U_{ij} \\ 0 & \text{if } X^o_{ij} > U_{ij} \end{cases},$$

for each $i=0,1,\ldots,k$. The observed values of (X_{ij},Δ_{ij}) will be denoted by (x_{ij},δ_{ij}). Thus, it is known which observations are uncensored (or due to the first cause of failure) and which ones are not. Also, the distribution of X_{ij} will be denoted by F_i and is given by

$$1-F_i = (1-F_{V_i})(1-H_i),\ i=0,1,\ldots,k.$$

Let $N = \sum_{i=0}^{k} n_i$ denote the total number of observations.

Before defining the estimator of F_{V_0}, a strongly consistent estimator of θ_i based on randomly right-censored data (Padgett and Wei, 1982) is discussed in Section 2. This estimator is then used in Section 3 to define a strongly consistent estimator of F_{V_0}. A numerical example illustrating the estimator of F_{V_0} in the case of two independent causes of failure is given in Section 4.

2. A CONSISTENT ESTIMATOR FOR θ_i

Let $x_{i1}, x_{i2},\ldots,x_{in_i}$ denote observed samples from F_i, $i=0,1\ldots,k$, respectively, where without loss of generality it is assumed that $x_{i1} \le x_{i2} \le\ldots\le x_{in_i}$ for each i. Padgett and Wei (1982) obtained a strongly consistent minimum distance estimator of the ratio of scale parameters in the two-sample problem based on randomly right-censored data. Their estimator was shown to perform better than other nonparametric estimators and will be used to estimate θ_i, the scale factor for F_{V_0} and F_{V_i}. This estimator will be defined next.

Let $y_{ij} = \log x_{ij}$, $j=1,\ldots,n_i$, $i=0,1,\ldots,k$, and let G_i denote the distribution function of $Y_{ij}^o = \log X_{ij}^o$. Define the Kaplan-Meier estimator of $G_i^c = 1-G_i$ by (Efron, 1967)

$$\hat{G}_i^c(y) = \begin{cases} 1, & y \le y_{i1} \\ \prod_{j=1}^{\ell-1} \left(\frac{n_i-j}{n_i-j+1}\right)^{\delta_{ij}}, & y \in (y_{i,\ell-1}, y_{i\ell}],\ \ell=2,\ldots,n_i \\ 0, & y > y_{in_i}, \end{cases}$$

for each $i=0,1,\ldots,k$. Also, define the jumps of $\hat{G}_i^c$ at the y_{ij}'s by a_{ij} (if $\delta_{ij} = 0$, the jump is zero), that is

$$a_{ij} = \begin{cases} \hat{G}_i^c(y_{ij}) - \hat{G}_i^c(y_{i,j+1}), & j=1,\ldots,n_i - 1 \\ \hat{G}_i^c(y_{in_i}), & j = n_i, \end{cases}$$

for each $i=0,\ldots,k$.

Let $\hat{\Delta}_{in_i}$ be a median of the discrete probability distribution with probability mass function given by

$$g_i(y) = \begin{cases} a_{0\ell}\, a_{ij}, \text{ if } & y=y_{0\ell} - y_{ij},\ \ell=1,\ldots,n_0,\ j=1,\ldots,n_i \\ 0, & \text{otherwise.} \end{cases}$$

Define $\hat{\theta}_{in_i} = \exp(\hat{\Delta}_{in_i})$. Then if $\sup\{x: H_i(x) < 1\} \ge \sup\{x: F_{V_i}(x) < 1\}$, $i=0,\ldots,k$, each $\hat{\theta}_{in_i}$ is a strongly consistent estimator of θ_i (Padgett and Wei, 1982). This result will be used to prove the strong consistency of the estimator proposed in Section 3.

3. A CONSISTENT ESTIMATOR OF F_{V_0}

For each $i=1,\ldots,k$, let $\hat{\theta}_{in_i}$ denote the strongly consistent estimator of θ_i defined in Section 2, and define $\hat{\theta}_{0n_0} \equiv 1$. The estimator of $1-F_{V_0}$ is then taken to be the Kaplan-Meier estimator $\hat{S}_0(t)$ obtained from the rescaled observations $Z_{ij} = \hat{\theta}_{in_i} X_{ij}$, $j=1,\ldots,n_i$, $i=0,\ldots,k$. Here it is assumed that the Z_{ij}'s are all distinct. If not, one of the methods given by Miller (1981) can be used to compute $\hat{S}_0(t)$. The strong consistency of $\hat{S}_0(t)$ as $n_i \to \infty$, $i=0,1,\ldots,k$, will now be proven. The following lemma will be needed for the proof.

Lemma. Let F be a continuous increasing function on $[0,\infty)$ such that $\lim_{x\to\infty} F(x) = c$, a finite constant. Let $i \in \{0,1,\ldots,k\}$ be fixed. Then there exists an event B_i which has probability one such that for $\omega \in B_i$ and a given $\varepsilon > 0$ there is an integer $N_1(i,\omega)$ for which $|F(x) - F(\theta_i x/\hat{\theta}_{in_i}(\omega))| < \varepsilon$ for all x whenever $n_i \geq N_1(i,\omega)$.

Proof. Let $i \geq 1$. By the strong consistency of $\hat{\theta}_{in_i}$, there is an event B_i which has probability one such that for each $\omega \in B_i$, $\hat{\theta}_{in_i}(\omega) \to \theta_i$ as $n_0, n_i \to \infty$.

Let $\varepsilon > 0$ and fix $\omega \in B_i$. By the hypotheses there exists an integer M such that $F(x) > c - \varepsilon$ for $x > M$. Let $\gamma > 0$. Then F is uniformly continuous on $[0, M+\gamma]$. Therefore, there is a $0 < \gamma_0 < \gamma$ so that if $|x - \theta_i x/\hat{\theta}_{in_i}(\omega)| < \gamma_0$, then $|F(x) - F(\theta_i x/\hat{\theta}_{in_i}(\omega))| < \varepsilon$. There is an integer $N_1(i,\omega)$ so that

$$|\theta_i/\hat{\theta}_{in_i}(\omega) - 1| < \gamma_0/(M+\gamma) \quad \text{for all } n_0, n_i \geq N_1(i,\omega),$$

(3.1) which implies that $|x - \theta_i x/\hat{\theta}_{in_i}(\omega)| < \gamma_0$ for all $n_0, n_i \geq N_1(i,\omega)$ whenever $x \in [0, M+\gamma]$.

Next, for $x > M + \gamma$, (3.1) implies $\theta_i x/\hat{\theta}_{in_i}(\omega) > M$ and the result follows by the choice of M. ///

The almost sure convergence of the estimator $\hat{S}_0(t)$ to $1 - F_{V_0}(t)$ can now be proven. For a distribution function G, define $T_G = \sup\{x: G(x) < 1\}$.

Theorem. $\hat{S}_0(t) \to S_0(t) = 1 - F_{V_0}(t)$ with probability one as $n_i \to \infty$, $i=0,1,\ldots,k$, for each $0 \le t \le t^*$, where $t^* = \min(T_{F_{V_0}}, T_{H_0})$.

Remark. The convergence of $\hat{S}_0$ holds only on $[0,t^*]$ since no observations larger than t^* can be obtained.

Proof. Define the subsurvival function estimators $\hat{R}^*_\ell(t)$ and $\hat{S}^*_\ell(t)$, $\ell=1,2$, by

$$\hat{R}^*_\ell(t) = \frac{1}{N}\sum_{i=0}^{k}\sum_{j=1}^{n_i} I[\theta_i X_{ij} > t,\ \delta_{ij} = \ell-1] = \frac{1}{N}\sum_{i=0}^{k} n_i \hat{R}^*_{\ell i}(t),$$

where

$$\hat{R}^*_{\ell i}(t) = \frac{1}{n_i}\sum_{j=1}^{n_i} I[\theta_i X_{ij} > t,\ \delta_{ij} = \ell-1],$$

and

$$\hat{S}^*_\ell(t) = \frac{1}{N}\sum_{i=0}^{k}\sum_{j=1}^{n_i} I[Z_{ij} > t,\ \delta_{ij} = \ell-1] = \frac{1}{N}\sum_{i=0}^{k} n_i\ \hat{S}^*_{\ell i}(t),$$

where

$$\hat{S}^*_{\ell i}(t) = \frac{1}{n_i}\sum_{j=1}^{n_i} I[Z_{ij} > t,\ \delta_{ij} = \ell-1]$$

and $I[\cdot]$ denotes the indicator function of the event enclosed in brackets. Note that $[Z_{ij} > t,\ \delta_{ij} = \ell-1]$ $= [\theta_i X_{ij} > \theta_i t/\hat{\theta}_{in_i},\ \delta_{ij} = \ell-1]$ and, thus, for each $i=0,1,\ldots,k$, $\hat{S}^*_{\ell i}(t) = \hat{R}^*_{\ell i}(\theta_i t/\hat{\theta}_{in_i})$, $\ell=1,2$, so that

$$\hat{S}_\ell^*(t) = \frac{1}{N}\sum_{i=0}^{k} n_i \hat{R}_{\ell i}^*(\theta_i t/\hat{\theta}_{in_i}), \ \ell=1,2.$$

To prove the result, we will show that

$$\hat{S}_\ell^*(t) \to S_\ell^*(t) \quad \text{uniformly in} \quad t \in [0,t^*]$$
$$\text{with probability one, } \ell=1,2, \tag{3.2}$$

where $S_\ell^*(t) = P(X_{0j} > t, \delta_{0j} = \ell-1)$, $\ell=1,2$. The result will then follow by Property 3.1 and Theorems 2.1 and 3.3 of Peterson (1977), since for all $0 \le t \le t^*$,

$$\hat{S}_0(t) = \Phi(\hat{S}_1^*(\cdot), \hat{S}_2^*(\cdot), t)$$
$$\to \Phi(S_1^*(\cdot), S_2^*(\cdot), t) = S_0(t) \quad \text{with probability one}$$

as $n_i \to \infty$, $i=0,1,\ldots,k$, where

$$\Phi(S_1^*(\cdot), S_2^*(\cdot), t) = \exp\left[\int_{S_1^*(-\infty)}^{S_1^*(t)} \frac{du}{u+S_2^*(S_1^{*-1}(u))}\right].$$

To prove (3.2), it is sufficient to show that for fixed i,

$$\hat{R}_{\ell i}^*(\theta_i t/\hat{\theta}_{in_i}) \to S_\ell^*(t) \quad \text{uniformly in} \quad t \in [0,t^*]$$
$$\text{with probability one, } \ell=1,2 \tag{3.3}$$

as $n_i \to \infty$, $i=0,1,\ldots,k$, since

$$|S_\ell^*(t) - \hat{S}_\ell^*(t)| = \left|\frac{1}{N}\sum_{i=1}^{k} n_i [S_\ell^*(t) - \hat{R}_{\ell i}^*(\theta_i t/\hat{\theta}_{in_i})]\right|.$$

Note that by Property 1 of Peterson (1977), $S_\ell^*(t)$, $\ell=1,2$, are continuous since $S_0 = 1-F_{V_0}$ and $1-H_0$ are. Let ℓ be fixed. By the Glivenko-Cantelli Lemma there exists an event A_i which has probability one such that for $\omega \in A_i$ and $\varepsilon > 0$ there is an integer $N_2(i,\omega)$ for which

$$|\hat{R}_{\ell i}^*(\theta_i t/\hat{\theta}_{in_i}(\omega)) - S_\ell^*(\theta_i t/\hat{\theta}_{in_i}(\omega))| < \varepsilon \tag{3.4}$$

whenever $n_0, n_i \ge N_2(i,\omega)$. Define the event $C_i = A_i \cap B_i$ where

B_i is the event defined in the Lemma. Let $\omega \in C_i$. Then for $n_0, n_i \ge N_0(i,\omega) = \max\{N_1(i,\omega), N_2(i,\omega)\}$,

$$\begin{aligned}|\hat{R}^*_{\ell i}(\theta_i t/\hat{\theta}_{in_i}(\omega)) - S^*_\ell(t)| \\ \le |\hat{R}^*_{\ell i}(\theta_i t/\hat{\theta}_{in_i}(\omega)) - S^*_\ell(\theta_i t/\hat{\theta}_{in_i}(\omega))| \\ + |S^*_\ell(\theta_i t/\hat{\theta}_{in_i}(\omega)) - S^*_\ell(t)| < 2\varepsilon,\ \ell=1,2,\end{aligned} \tag{3.5}$$

by the Lemma (applied to $1-S^*_\ell$) and (3.4). Finally, (3.4) and (3.5) hold for all $i=0,1,\ldots,k$ whenever $\omega \in C = \bigcap_{m=1}^{k} C_m$ and $n_i \ge \max\{N_0(m,\omega),\ m=1,\ldots,k\}$ for all $i=0,1,\ldots,k$. Therefore, the result (3.3) follows since the event C has probability one.

///

A robust estimator of the mean life under the normal stress V_0 can be obtained by (see, for example, Miller, 1981)

$$\hat{\mu}_0 = \int_0^\infty \hat{S}_0(t)dt. \tag{3.6}$$

Since (3.6) can be written $\int_0^{t^*} \hat{S}_0(t)dt + \int_{t^*}^\infty \hat{S}_0(t)dt$, the strong consistency of $\hat{\mu}_0$ follows by the bounded convergence theorem whenever $t^* < \infty$. A rough asymptotic estimate of the standard error of $\hat{\mu}_0$ can be computed from the formula on page 71 of Miller (1981), but since $\hat{S}_0$ involves only estimates of the θ_i's, this standard error estimate is not very reliable.

4. EXAMPLE

As a numerical example, we use data from accelerated life tests on Buna N rubber O-rings which are used as seals in components of nuclear reactors (adapted from Carfagno and Gibson, 1980). Life tests were conducted on 30 O-rings, with 10 O-rings at the normal operating temperature of 200°F and ten at each of the two accelerated temperatures of 250°F and 275°F with no other stress present. Failure (leakage) of the O-rings could be attributed to either radial fracture or circumferential cracks usually on the outside diameter. Hence, there are two causes of failure, assumed to be independent. The data for this example are shown in Table 1, where "+" denotes an observed failure due to circumferential cracks.

Table 1. O-Ring Failure Times in Hours

200°F	250°F	275°F
1683.08	399.56	147.23
1784.32	422.40+	147.90
1683.50+	333.07	128.87+
1784.49	444.72+	147.86
1862.32+	367.80	165.13
1784.84	367.49	147.20
1683.16	393.55	147.16
1784.70+	393.10	165.82+
1683.71+	314.63	147.89
1784.23	292.88	147.79

It is desired to estimate the life distribution for the more common failure cause, radial fracture, as well as the mean life μ_0 for this type of failure at the normal operating temperature of 200°F. Here $n_i = 10$, $i=0,1,2$, and all $\delta_{ij} = 1$ except $\delta_{03}, \delta_{05}, \delta_{08}, \delta_{09}, \delta_{12}, \delta_{14}, \delta_{23}, \delta_{28}$, which are zero. The computation of the estimates of the scale factors yields $\hat{\theta}_{1,10} = 4.6964$ and $\hat{\theta}_{2,10} = 12.0678$ (recall that $\hat{\theta}_{0,10} \equiv 1$). Hence, the rescaled observed values are $z_{ij} = \hat{\theta}_{i,10}\, x_{ij}$, $j=1,\ldots,10$ and $i=0,1,2$. Let $z_{(m)}$, $m=1,2,\ldots,N$, denote the $N=30$ ordered observed z_{ij}'s. The ordered observations along with the corresponding values of the Kaplan-Meier estimate $\hat{F}_{V_0}(t) = 1-\hat{S}_0(t)$ of $F_{V_0}(t)$ at $t = z_{(m)}$, $m=1,\ldots,N$ are shown in Table 2 and Figure 1. The estimated mean life is obtained from Table 2 by

$$\hat{\mu}_0 = \int_0^\infty \hat{S}_0(t)\,dt = \sum_{m=1}^{N} \hat{S}_0(z_{(m)})(z_{(m)} - z_{(m-1)}) = 1804.14 \text{ hours},$$

where $z_{(0)} \equiv 0$. A rough estimate of the standard deviation of $\hat{\mu}_0$ is obtained from (Miller, 1981)

$$\left\{ \sum_{\{m: z_{(m)} \text{ is uncensored}\}} \left[\sum_{j=m}^{N-1} \hat{S}_0(z_{(j)})(z_{(j+1)} - z_{(j)}) \right]^2 \times \frac{1}{(N-m+1)(N-m)} \right\}^{\frac{1}{2}} = 60.92 \text{ hours.}$$

Table 2. Ordered Rescaled Observations and $\hat{F}_{V_0}$

$z_{(m)}$	0	1375.48	1477.62	1555.18+	1564.22
$\hat{F}_{V_0}(z_{(m)})$	0	0.0333	0.0667	0.0667	0.1012
$z_{(m)}$	1683.08	1683.16	1683.50+	1683.71+	1725.87
$\hat{F}_{V_0}(z_{(m)})$	0.1358	0.1704	0.1704	0.1704	0.2081
$z_{(m)}$	1727.33	1775.90	1776.38	1776.75	1783.50
$\hat{F}_{V_0}(z_{(m)})$	0.2458	0.2835	0.3212	0.3589	0.3966
$z_{(m)}$	1784.23	1784.32	1784.35	1784.49	1784.70+
$\hat{F}_{V_0}(z_{(m)})$	0.4343	0.4721	0.5098	0.5475	0.5475
$z_{(m)}$	1784.71	1784.83	1784.84	1846.15	1848.26
$\hat{F}_{V_0}(z_{(m)})$	0.5886	0.6298	0.6709	0.7120	0.7532
$z_{(m)}$	1862.32+	1876.49	1983.75+	1992.76	2001.09+
$\hat{F}_{V_0}(z_{(m)})$	0.7532	0.8025	0.8025	0.8684	0.8684
$z_{(m)}$	2008.58+				
$\hat{F}_{V_0}(z_{(m)})$	1.0000				

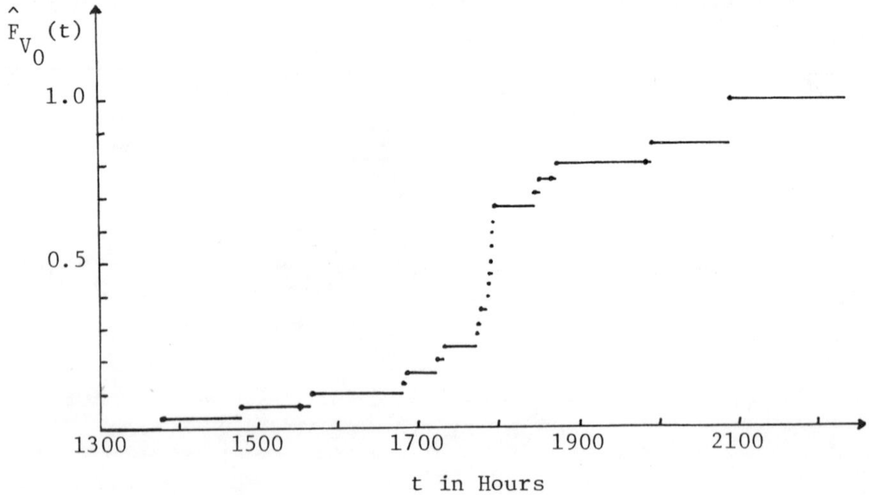

Figure 1. Estimate of F_{V_0}.

5. SUMMARY

We have proposed a nonparametric estimator of the nonaccelerated life distribution from accelerated life test data. The data may be randomly censored from the right at each stress level or be observed under two independent causes of failure at each stress level. The estimator did not require the assumption of an acceleration model, but assumed that the accelerated distributions differed from the nonaccelerated distribution only by a scale factor. Also, as with some previous nonparametric procedures for accelerated life tests, observations from the nonaccelerated environment were needed. It was shown that the estimator is strongly consistent under general conditions. In addition, a strongly consistent robust estimator of mean life was easily obtained.

REFERENCES

Barlow, R. E., and Scheuer, E. M. (1971), "Estimation from Accelerated Life Tests," *Technometrics*, 13, 145-149.

Basu, A. P., and Ebrahimi, N. (1982), "Nonparametric Accelerated Life Testing," *IEEE Trans. Reliability*, R-31, 432-435.

Carfagno, S. P., and Gibson, R. J. (1980), *A Review of Equipment Aging Theory and Technology*, NP-1558, Research Project 890-1, Franklin Research Center, Philadelphia, Pa.

Efron, B. (1967), "The Two Sample Problem with Censored Data," *Proc. 5th Berkeley Sympos. Math. Statist. Prob.*, 4, 831-853.

Miller, R. G. (1981), *Survival Analysis*, New York: John Wiley.

Padgett, W. J., and Wei, L. J. (1982), "Estimation of the Ratio of Scale Parameters in the Two Sample Problem with Arbitrary Right Censorship," *Biometrika*, 69, 252-256.

Peterson, A. V., Jr. (1977), "Expressing the Kaplan-Meier Estimation as a Function of Empirical Subsurvival Functions," *Journal of the American Statistical Association*, 72, 854-858.

Shaked, M., and Singpurwalla, N. D. (1982), "Nonparametric Estimation and Goodness of Fit Testing of Hypothesis for Distribution in Accelerated Life Testing," *IEEE Trans. Reliability*, R-31, 69-74.

Shaked, M., Zimmer, W. J., and Ball, C.A. (1979), "A Nonparametric Approach to Accelerated Life Testing," *Journal of the American Statistical Association*, 74, 694-698.

Steck, G. P., Zimmer, W. J., and Williams, R. E. (1974), "Estimation of Parameters in Accelerated Models," *Proc. 1974 Annual Reliability and Maintainability Symposium, IEEE*, Piscataway, N.J., 428-431.

A KALMAN FILTER APPROACH TO ACCELERATED LIFE TESTING--A PRELIMINARY DEVELOPMENT

Richard J. Meinhold
Nozer D. Singpurwalla[1]

Department of Operations Research
and Department of Statistics
The George Washington University
Washington, D.C.

The Kalman filter model, successfully used in a variety of situations, can also be used for inference from accelerated life tests. The use of this model calls for the specification of the "law of motion," and this means that a time transformation function such as the Arrhenius law, the power law, etc., must be specified. The ordered values of the stresses can be conceptualized as the ordered values of time, and thus inference about the life behavior at low stress can be viewed as inference about the "state of nature," at a future time. The Kalman filter model has a fully Bayesian interpretation, and thus its use in accelerated life testing makes inference from such tests properly Bayesian and therefore coherent. This paper is preliminary, and we intend to present here the feasibility of such an approach.

[1]*Research supported by the Army Research Office, Grant DAAG 29-83-K-0013, and the Office of Naval Research Contract N00014-75-C-0263, Project NR-042-372.*

ISBN 0-12-041420-1

I. THE KALMAN FILTER MODEL—AN INTRODUCTION

Let $Y_t, Y_{t-1}, \ldots, Y_1$, denote the observed values of a variable of interest at times t, $t-1$, ..., 1. Suppose that Y_t depends on an unobservable quantity θ_t, the *state of nature* about which we wish to make inferences. The relationship between Y_t and θ_t is linear, and is specified by the *observation equation*

$$Y_t = F_t\theta_t + v_t \ , \tag{1.1}$$

where F_t is a known quantity. The *observation error* v_t is assumed to be normally distributed with mean zero and a known variance V_t, denoted as $v_t \sim N(0,V_t)$. The state of nature θ_t is dynamic and changes with time according to the *system equation*

$$\theta_t = G_t\theta_{t-1} + w_t \ , \tag{1.2}$$

where G_t is a known quantity, and $w_t \sim N(0,W_t)$, with W_t also known. Even though the quantities F_t, G_t, V_t, and W_t are indexed by t, implying that they change with time, they may in some applications be assumed to be constant. It is often convenient to assume that v_t is independent of w_t.

The term "Kalman filter" or "Kalman filtering" refers to a recursive procedure for inference about the state of nature θ_t. The key notion here is that, given the data $\underset{\sim}{Y}_t = (Y_t, \ldots, Y_1)$, inference about θ_t can be carried out through a direct application of Bayes' theorem:

$$P(\theta_t|\underset{\sim}{Y}_t) \propto P(Y_t|\theta_t, \underset{\sim}{Y}_{t-1}) \times P(\theta_t|\underset{\sim}{Y}_{t-1}) \ .$$

Note that the expression on the left denotes the posterior distribution for θ at time t, whereas the first and second expressions on the right-hand side denote the likelihood and the prior distribution for θ, respectively. Details of the recursive procedure are given in Meinhold and Singpurwalla, 1982.

II. APPLICATION TO ACCELERATED LIFE TESTING

The Kalman filter model can be used for making inference from accelerated life tests and other dose-response experiments whenever the use of certain time transformation functions can be physically justified. To see this, suppose that T, the time to failure random variable, has an exponential distribution with mean $\theta > 0$; that is,

$$f(t;\theta) = \frac{1}{\theta} \exp(-t/\theta), \qquad t \geqslant 0,\ \theta > 0$$

$$= 0 \qquad , \qquad \text{otherwise.}$$

Suppose that $S_1 > S_2 > \ldots > S_t > S_{t+1}$ are stress levels of decreasing intensity, which may or may not be equally spaced. The use conditions stress at which it is very difficult and/or time-consuming to obtain failure data is S_{t+1}, and our goal is to make inferences about θ_{t+1}, the mean time to failure under S_{t+1}. In general, θ_j denotes the mean time to failure under S_j, $j = 1,2,\ldots, t+1$.

The following physical models which describe the relationship between θ_j and S_j have been often used in practice (see, for example, Mann, Schafer, and Singpurwalla, 1974, p. 421). These relationships give us the system equations needed for setting up the Kalman filter model. They are:

(a) the Arrhenius law, $\theta_t = \exp(\alpha + \phi_t/S_t)$, and (2.1)

(b) the Power Law, $\theta_t = \alpha S_t^{-\phi_t}$, (2.2)

where α and ϕ_t are unknown constants. For a recent discussion involving the use of these models, we refer the reader to Sethuraman and Singpurwalla, 1982, and Shaked and Singpurwalla, 1982,1983. On the basis of the above, we *postulate* the system equations

$$\log \theta_t = \log \theta_{t-1} + \phi_t\, h(S_t,S_{t-1}) + u_t \qquad (2.3)$$

and

$$\phi_t = \phi_{t-1} + v_t, \tag{2.4}$$

with

$$\begin{bmatrix} u_t \\ v_t \end{bmatrix} \sim N[\underline{0}, \Sigma_t];$$

$h(S_t, S_{t-1})$ is a function of S_t and S_{t-1}. The notation $(X_1, X_2)' \sim N[\underset{\sim}{\mu}, \Sigma]$ denotes the fact that the column vector $(X_1, X_2)'$ has a multivariate normal distribution with mean vector $\underset{\sim}{\mu}$ and covariance matrix Σ.

For the Arrhenius law, the function $h(S_t, S_{t-1})$ turns out to be $((1/S_t) - (1/S_{t-1}))$. To see this, we eliminate α in (2.1) to obtain

$$\begin{aligned} \log \theta_t &= \log \theta_{t-1} + \frac{\phi_t}{S_t} - \frac{\phi_{t-1}}{S_{t-1}} \\ &= \log \theta_{t-1} + \frac{\phi_t}{S_t} - \left[\frac{\phi_t}{S_{t-1}} - \frac{v_t}{S_{t-1}}\right] \\ &= \log \theta_{t-1} + \phi_t\left[\frac{1}{S_t} - \frac{1}{S_{t-1}}\right] + \frac{v_t}{S_{t-1}} \end{aligned} \tag{2.5}$$

For the power law, the function $h(S_t, S_{t-1})$ is $\log(S_{t-1}/S_t)$, and this too can be verified by eliminating the α in (2.2) to obtain

$$\log \theta_t = \log \theta_{t-1} + \phi_t \log\left(\frac{S_{t-1}}{S_t}\right) + v_t \log\left(\frac{1}{S_{t-1}}\right). \tag{2.6}$$

If we compare (2.5) and (2.6) with the system equation (2.3), we note that the error u_t becomes v_t/S_{t-1} and $v_t \log(1/S_{t-1})$, respectively. Thus, in our formulation of the Kalman filter model, the error of the second system equation is a component of the error of the first system equation.

In view of the above, if we let $u_t = w_t + \beta_t v_t$, where β_t is a known function of S_{t-1}, and w_t is an innovation which is independent of v_t, then we may determine the structure of Σ_t, where

$$\Sigma_t = \begin{bmatrix} \mathrm{Var}(w_t) + \beta_t^2 \,\mathrm{Var}(v_t) & \beta_t \,\mathrm{Var}(v_t) \\ \beta_t \,\mathrm{Var}(v_t) & \mathrm{Var}(v_t) \end{bmatrix}$$

by specifying $\mathrm{Var}(w_t)$ and $\mathrm{Var}(v_t)$. To initiate the recursive filtering procedure, we will also have to specify θ_0 and ϕ_0.

A. *Distribution Theory for the Observation Equation*

Having settled upon the system equations (2.3) and (2.4), our next task is to formulate an observation equation which is in conformity with the assumptions following (1.1). To do this, suppose that $\tau_{j1}, \tau_{j2}, \ldots, \tau_{jn_j}$ are the times to failure of the n_j items tested under stress S_j, $j = 1,\ldots,t$. For the purpose of this discussion we shall assume no censoring and/or withdrawal, so that we have complete samples. Suppose that the outcomes of the tests at the various stress levels are summarized by the sample mean times to failure $X_j = (n_j)^{-1} \sum_{i=1}^{n_j} \tau_{ji}$, $j = 1,\ldots,t$. Then X_j will have a gamma distribution with parameters n_j and θ_j/n_j, and $Y_j = -\log X_j$ will have the extreme value distribution with parameters n_j and θ_j. That is, the probability density function of Y_j is

$$g(y;n_j,\theta_j) = \frac{n_j^{n_j}}{\Gamma(n_j)\theta_j^{n_j}} \exp\{-n_j y - \frac{n_j}{\theta_j} \exp(-y)\}. \tag{2.7}$$

The moment generating function of the above density is, for $s < n_j$

$$M(s) = \frac{\Gamma(n_j - s)}{(n_j)} \left(\frac{n_j}{\theta_j}\right)^s$$

and $K(s) = \log M(s)$, the cumulant generating function, is

$$K(s) = s(\log n_j - \log \theta_j) + \log \Gamma(n_j - s) - \log \Gamma(n_j).$$

The cumulant generating function yields the cumulants

$$K_1 = \log n_j - \log \theta_j + \psi(n_j),$$

and

$$K_r = (-1)^r \psi^{(r-1)}(n_j) \qquad , \qquad r \geqslant 2,$$

where $\psi^{(m+2)}(\cdot)$ is the m-gamma function.

It is instructive to examine the cumulants for two reasons. First, we note that $\text{Var}(Y_j) = K_2 = \psi^1(n_j)$, and so the variance of the log transformed variable depends only on the known quantity n_j, and not on θ_j. Thus the log transformation has stabilized the variance. Secondly, the measures of skewness and kurtosis of the above distribution, as determined by the cumulants, approach those of the normal distribution more rapidly, as $n_j \to \infty$, than those of the original gamma random variables.

In view of the above, if we therefore define $Z_j = \log n_j + \psi(n_j) - Y_j$, we may then hypothesize that for sufficiently large n_j, Z_j is approximately normal with mean $\log \theta_j$ and variance $\psi^1(n_j)$. We now have as our observation equation

$$Z_j = \log \theta_j + v_j,$$

where $v_j \simeq N(0, \psi^1(n_j))$, and $\psi^1(n_j)$ is known.

From this point onward, the application of the Kalman filter is mechanical. When the underlying distribution of life lengths is not exponential, the distribution of the mean lifetimes X_j, for small values of n_j, is complicated and requires further analysis—this is currently in progress. Approaches for using the Kalman filter model when the distribution of life lengths cannot be specified are also currently under investigation.

REFERENCES

Mann, N. R., R. E. Schafer, and N. D. Singpurwalla. *Methods for Statistical Analysis of Reliability and Life Data.* Wiley, New York (1974).

Meinhold, R. J., and N. D. Singpurwalla. "Understanding the Kalman Filter." *Amer. Statistician*, *37*, 2, pp. 123-127 (1982).

Sethuraman, J., and N. D. Singpurwalla. "Testing of Hypotheses for Distributions in Accelerated Life Testing." *J. Amer. Statist. Assoc.*, *77*, 377, pp. 204-208 (1982).

Shaked, M., and N. D. Singpurwalla. "Nonparametric Estimation and Goodness of Fit Testing of Hypotheses for Distribution in Accelerated Life Testing." *IEEE Trans. Reliability*, *R-31*, 1, pp. 69-74 (1982).

Shaked, M., and N. D. Singpurwalla. "Inference for Step-stress Accelerated Tests." *J. Statist. Planning & Inference*, *7*, pp. 295-306 (1983).

INFERENCE FROM ACCELERATED LIFE TESTS

W. J. Padgett

Department of Mathematics and Statistics
University of South Carolina
Columbia, South Carolina

This paper is intended to be a brief exposition of the basic ideas of inference from accelerated life tests. Many of the recent parametric and nonparametric approaches are outlined, including results based on censored as well as complete samples from accelerated tests.

1. INTRODUCTION

Many devices such as electronic items have very high reliability when operating within their intended normal use environment. This presents problems in measuring the reliability of such devices because a very long period of testing under the actual operating conditions would be required to obtain sufficient data to estimate the reliability. Even if this testing could be accomplished, the time frame is such that the devices may become obsolete before their reliability is established due to the high rate of technological advances. Also, it would be difficult to conduct the testing in laboratory conditions that would closely approximate the normal use environment and hold those conditions essentially constant over the long period of testing, possibly many years.

One solution to the problem of obtaining meaningful life test data for high reliability devices is *accelerated life testing*. This type of testing involves observing the performance of these kinds of devices operating at higher stress levels than usual to obtain failures more quickly. In order to shorten product life, it is a well-established engineering practice to use certain stresses or accelerating variables, such as higher levels of temperature, voltage, pressure, vibration, etc., than the normal operating level.

The main difficulty of accelerated life testing lies in using the failure data obtained at the accelerated, or higher stress, conditions to predict the reliability, mean life, or other quantities under the normal use condition. Extrapolation from the accelerated stresses to the normal use stress is done by choosing an appropriate model, called an *acceleration model*. The choice of an acceleration model calls for a knowledge of the variation of failure behavior with environment. In parametric methods of accelerated life testing, this involves functional relationships between the parameters of the failure distributions and the environmental stresses. The functional relationships may also involve unknown parameters. In nonparametric approaches, where no specific form of the failure distributions is specified, the change in the failure distribution due to a change in environmental stress is often assumed to be a simple scale change or a time transformation which involves unknown parameters. In either the parametric or nonparametric approach, all unknown parameters must be estimated from the accelerated test data in order to extrapolate to the normal use stress.

Various acceleration models may be derived from a knowledge of the physics of failure of the device under test. Four acceleration models which are commonly used are the power rule model, the Arrhenius model, the Eyring model, and the generalized Eyring model. These models will be discussed briefly in Section 3. Mann, Schafer, and Singpurwalla (1974) contains a description of them.

This paper is not intended to be a review of the area of accelerated life testing, but will contain only an exposition of various inference procedures from accelerated test data. In Sections 4 and 5 some of the representative results that have been obtained in the parametric, partially nonparametric, and nonparametric settings will be mentioned briefly. There are several excellent books and articles which discuss methods for planning and analyzing accelerated tests. Among them are Mann, Schafer, and Singpurwalla (1974, Ch. 9), Little and Jebe (1974), Kalbfleisch and Prentice (1980), Carfagno and Gibson (1980), and Nelson (1974). Meeker (1983a) gives an extensive bibliography on accelerated life tests, including specific applications, and a review of statistical aspects of accelerated testing is given by Meeker (1983b).

2. NOTATION

In this section the notation that will be used throughout the remainder of this paper will be introduced. Since inference from accelerated life tests has been studied for both complete samples and censored samples under various censoring mechanisms, the concepts of censoring will also be discussed briefly.

Let F_V denote the life distribution of items on test under stress level V. Let $V_1,\ldots,V_k$ be k stress levels under accelerated testing and let V_0 denote the normal use stress. It usually will be assumed that the V_i's are in increasing order of stresses so that V_0 will tend to produce much fewer failures than will V_k over the same time period. The choice of k and V_i's is a problem in the design and planning of accelerated life tests and will be discussed later (see Mann, Schafer, and Singpurwalla, 1974, pp. 456-461; Escobar and Meeker, 1983; Meeker, 1984; Nelson, 1974; Nelson and Kielpinski, 1976; and Nelson and

Meeker, 1978, for example).

Let X_i^0 denote the true lifetime random variable at stress level V_i, i=0,1,2,...,k. The probability distribution of X_i^0 is F_{V_i}. If n_i randomly chosen items are placed on life test at stress level V_i, their lifetimes will be denoted by X_{ij}^0, $j=1,2,\ldots,n_i$, for each $i=0,1,\ldots,k$. The total number of items put on test is $N = \sum_{i=0}^{k} n_i$. If the life tests are continued until all n_i items fail at the stress level V_i, then the observed failure times will be denoted by $X_{i1}^0,\ldots,X_{in_i}^0$ for each i. In this case X_{ij}^0, $j=1,\ldots,n_i$, will be called a *complete sample* of failure times at stress V_i. (It might be that no items are placed on test at the normal use stress V_0).

In many accelerated life testing situations, some items at one or more of the stress levels may be removed from test at various times due to circumstances beyond the experimenter's control, or some of the tests at the various levels of stress may be terminated before all items fail. In these circumstances a *censored sample* is observed. The various types of censorship that are encountered in practice may be defined as special cases of the more general arbitrary right-censorship. To define an *arbitrarily right-censored sample,* it is assumed that under stress V_i there is a sequence of constants or random variables $U_{i1},\ldots,U_{in_i}$ which censors $X_{i1}^0,\ldots,X_{in_i}^0$, respectively, from the right. That is, the observations available to the experimenter are the pairs (X_{ij},Δ_{ij}), $j=1,2,\ldots,n_i$, where

$$X_{ij} = \min\{X_{ij}^0,U_{ij}\}, \quad \Delta_{ij} = \begin{cases} 1, & X_{ij}^0 \le U_{ij} \\ 0, & X_{ij}^0 > U_{ij} \end{cases} .$$

Thus, the experimenter knows which X_{ij}'s are right-censored and which ones are not. The type of censorship depends on the nature of the U_{ij}'s: (i) If the U_{ij}'s are all constants, then the X_{ij}'s are *time-censored;* if $U_{ij} = c_i$, a constant for all

$j=1,\ldots,n_i$, then the observed sample at stress level V_i is said to be a *Type I censored sample*. (ii) If $U_{ij} = X^0_{i(r_i)}$, the r_ith order statistic among $X^0_{i1},\ldots,X^0_{in_i}$ for all $j=1,\ldots,n_i$, then the sample is a simple *Type II censored sample*. That is, the life test at stress level V_i is terminated after r_i failures have occurred. (iii) If for each i, the U_{ij}'s are independent, identically distributed random variables having common distribution function H_i and independent of $X^0_{i1},\ldots,X^0_{in_i}$, then the observed sample (X_{ij},Δ_{ij}), $j=1,\ldots,n_i$, is a *randomly right-censored sample*. In this case, it is also usually assumed that the censoring random variables $U_{i1},\ldots,U_{in_i}$, $i=0,1,\ldots,k$, are independent; that is, censoring at one stress level has no effect on the censoring at other stress levels.

In the case of random right-censorship (iii), the random variables X^0_{ij} and U_{ij} for fixed i and j can be considered as the lifetimes of the jth item on test at stress level V_i due to two independent causes of failure, or independent competing risks. The lifetime variable X^0_{ij} denotes the lifetime due to the "main" cause of failure, or the failure mode of interest, and U_{ij} denotes the lifetime due to the other independent competing cause. This situation has been extended to more than two independent competing risks by Basu and Ebrahimi (1982).

3. ACCELERATION MODELS

The use of accelerated life testing to make inferences about the normal use life distribution requires a model to relate the life length to the stress levels that are to be applied to the items being tested. This model is referred to as the *acceleration model*. Its selection is probably the most serious difficulty in accelerated testing procedures. The acceleration model should be physically reasonable for the particular type of item or product being tested and the kind of stress being applied to accelerate failure. As Meeker (1983a) points out, whenever possible, an acceleration model should be empirically verified for the particular product under test.

In this section some of the various acceleration models that have been used in parametric and nonparametric inference procedures for accelerated life tests will be described briefly. A detailed discussion of the power rule model, the Arrhenius model, and the Eyring model can be found in Mann, Schafer, and Singpurwalla (1974, pp. 420-422).

In a general parametric setting, suppose the lifetime random variable X_i^0 of items in an environment described by a constant stress level V_i has a probability distribution $F^0(t;\underline{\theta}_i)$ depending on a vector of parameters $\underline{\theta}_i$. Note that, more generally, a vector of stresses $\underline{V}_i$ may be applied to items on test, but for simplicity, in this paper a single stress V_i will be assumed for each of the k accelerated environments, as well as for the normal use environment. Two assumptions which are made (Mann, Schafer, and Singpurwalla, 1974) are: (1) The change in stress level does not change the *type* of the lifetime distribution $F^0(t;\underline{\theta})$, but changes only the parameter values, and (2) the relationship between the stress level V and the parameters $\underline{\theta}$, say $\underline{\theta} = m(V;\alpha,\beta,\ldots)$, is known except for one or more of the acceleration parameters $\alpha,\beta,\ldots,$ and that the relationship holds for a certain range of values of V. In this situation, the goal is to obtain estimates of the unknown $\alpha,\beta,\ldots$ based on life test data obtained at large values of V and make inferences about $\underline{\theta}$ for the normal use stress V_0 through the assumed acceleration model $\underline{\theta} = m(V;\alpha,\beta,\ldots)$.

The exponential distribution with parameter λ is widely used as a lifetime distribution, so the acceleration models will be discussed here for exponential distributions. Several authors have considered other lifetime distributions such as Weibull (Mann, 1972, and Nelson, 1975), extreme value (Meeker and Nelson,

1975, and Nelson and Meeker, 1978), and lognormal (Nelson and Kielpinski, 1976), for example. It will be assumed that under constant application of a single stress at level V_i, the item being tested has an exponential lifetime distribution with mean μ_i and probability density function $f^0(t;\lambda_i) = \lambda_i \exp(-\lambda_i t)$, $t \geq 0$, $\lambda_i > 0$, and $=0$, otherwise. Then $\mu_i = 1/\lambda_i$ is the mean time to failure under stress level V_i. The following acceleration models (relationships between λ_i and V_i) have been suggested in the literature.

The *power rule (or inverse power) model* can be derived by considerations of kinetic theory and activation energy. This model has applications to fatigue testing of metals, the dielectric breakdown of capacitors, and aging of multicomponent systems. The model is $\mu_i = \alpha V_i^{-\beta}$, where $\alpha > 0$ and $\beta > 0$ are unknown parameters. It is desirable to estimate α and β from life test data at stress levels $V_1,\ldots,V_k$ and make inferences about $\mu_0 = 1/\lambda_0$ at the normal use stress V_0.

The *Arrhenius model* expresses the degradation rate of a parameter of the device as a function of its operating temperature. It is usually applied to thermal aging and is applicable to semiconductor materials. The relationship is $\lambda_i = \exp(\alpha-\beta/V_i)$, where V_i denotes the temperature stress level and α and β are unknown parameters to be estimated in order to make inferences about λ_0 at normal temperature level V_0.

The *Eyring model* can be derived from principles of quantum mechanics and, generally, has application to accelerated testing of devices subjected to a constant application of two types of stresses, one thermal and one nonthermal. The model is $\lambda_i = \alpha T_i \exp(-\beta/KT_i)\exp(\gamma V_i + \delta V_i/KT_i)$, where α,β,γ, and δ are unknown parameters to be estimated, K denotes Boltzmann's constant (1.38×10^{-16} erg/degree Kelvin), and T_i is the thermal stress level and V_i is the nonthermal stress. In the absence of a nonthermal stress, the *Eyring model* reduces to $\lambda_i = \alpha T_i \exp(-\beta/T_i)$.

Chernoff (1962) considered an acceleration model for exponential lifetimes with means $\mu_i = (\alpha V_i + \beta V_i^2)^{-1}$, where $\alpha > 0$ and $\beta > 0$ were unknown parameters. Thus λ_i was a quadratic function of the stress level. Chernoff also considered models for three-dimensional vector stresses $\underline{V}_i = (V_{1i}, V_{2i}, V_{3i})$.

In partially nonparametric approaches to inference from accelerated life tests, no particular form of the lifetime distribution is assumed, but an acceleration model is used (see, for example, Shaked, Zimmer, and Ball, 1979; Sethuraman and Singpurwalla, 1982; Shaked and Singpurwalla, 1982; Basu and Ebrahimi, 1982; and Shaked and Singpurwalla, 1983). Shaked, Zimmer, and Ball (1979) assumed that the k accelerated stress levels $V_1, \ldots, V_k$ were selected and fixed at the outset and that for every pair of stresses V_i, V_j, $i,j = 0,1,\ldots,k$, a known function m existed so that the lifetime distributions satisfied $F_{V_j}(t) = F_{V_i}(m(\alpha, V_j, V_i, t))$, $t \geq 0$, where α was an unknown parameter (or vector of parameters). The form of F_{V_i} was not assumed to be known. Various choices of m gave the usual power rule, Arrhenius, Eyring, etc., acceleration models. The other references mentioned above assume acceleration models which can be obtained as special cases of m.

In a totally nonparametric setting, there is no assumption made about the form of the underlying lifetime distribution at the various stress levels nor about the form of an acceleration model. In this setting, the acceleration assumption is typically that the life distributions are stochastically ordered with respect to increasing levels of stress (Barlow and Scheuer, 1971) or that the lifetime distributions at two distinct stress levels differ only by a scale change (Louis, 1981; Spurrier, 1983; Wei and Gail, 1983; McNichols and Padgett, 1984). For these procedures, usually it must be assumed that failure data are available from the normal use stress as well as from accelerated stresses.

Other approaches to acceleration models include Bayesian

techniques (Proschan and Singpurwalla, 1980), the Kalman filter (Meinhold and Singpurwalla, 1984), and the ideas of shock models and wear processes (Shaked and Singpurwalla, 1983).

In most of the acceleration models mentioned in this section, the requirement is that the stress level V be constant for an item until it fails or is withdrawn from the test. The exception to this is the situation given by Shaked and Singpurwalla (1983). Other references which drop this requirement are De Groot and Goel (1979) and Nelson (1980). In addition, Allen (1959) allowed the stress level to be a continuously increasing function of time.

The *design* aspects of accelerated life testing experiments involve the selection of stress levels, the number of stress levels, and the number of items to be tested at each stress level. A well-designed accelerated test plan can result in improved precision of estimators and allow for censoring. Optimum design of accelerated tests selects the stress levels and allocation of items to be tested to satisfy some optimality criteria. Mann, Schafer, and Singpurwalla (1974, Sec. 9.6) discuss accelerated testing as a factorial experiment, with the stresses as the factors in the design. Meeker (1984) compares optimum accelerated life test plans and some compromise test plans for Weibull and lognormal distributions under Type I censoring. He compares in particular the ability of the plans to detect departure from the assumed acceleration model. Escobar and Meeker (1983) give methods for planning accelerated life tests for models in which time to failure has a location-scale distribution and the location parameter can be expressed as a function of the stress level. They allowed Type II censored data. Other accelerated testing plans were developed by Chernoff (1962), Hahn and Meeker (1978), Mann (1972), Meeker and Nelson (1975), Nelson and Kielpinski (1975, 1976), and Nelson (1978), among others.

4. SOME PARAMETRIC PROCEDURES

Inference from accelerated life tests can become quite involved, even when the functional relationship between the stress level and the single parameter of the life distribution is simple. This will be illustrated in the present section with the simplest parametric case of an exponential life distribution at each stress level and the power rule model as the acceleration model. These results are discussed by Mann, Schafer, and Singpurwalla (1974, Sect. 9.2-9.3) and Singpurwalla (1971). Also, in this section some results of Shaked (1978) will be stated that generalize the power rule model and Arrhenius model with life distributions which have hazard rate functions of the form $r(t) = Ag(t) + Bh(t)$. This class of life distributions contains the exponential distributions as well as other distributions which have been considered in the literature. Results for Weibull and other life distributions can be found, for example, in papers by Mann (1972), Nelson (1972, 1975), and Nelson and Meeker (1978), and will not be discussed in this paper.

Suppose that life tests are to be conducted at k accelerated stress levels, V_i, $i=1,\ldots,k$, on devices which are assumed to have an exponential life distribution with parameter $\lambda_i = 1/\mu_i$, where μ_i is the mean lifetime under stress V_i. In order to assure an independence among the k life tests, a randomization procedure is performed as follows (Mann, Schafer, and Singpurwalla, 1974, p. 423): Choose a value V_j at random from among the k stresses V_i, $i=1,\ldots,k$, and put n_j devices on life test under constant application of stress V_j. The test is terminated after r_j failures have occurred (Type II censoring) and the times to failure, $t_{1j}, t_{2j},\ldots,t_{r_j j}$ are recorded. Next choose another stress V_ℓ at random from the remaining $k-1$ stress levels and repeat the procedure. Continue this procedure until all k life tests have been performed, yielding as data the set $\{V_i, n_i, r_i, \hat{\mu}_i\}$, $i=1,\ldots,k$, where $\hat{\mu}_i$ is an estimator of μ_i. The unique minimum variance unbiased estimator of μ_i is (Epstein and Sobel, 1954)

$$\hat{\mu}_i = r_i^{-1}\left[\sum_{j=1}^{r_i} t_{ji} + (n_i - r_i)t_{r_i i}\right],$$

which is known to have a gamma distribution.

Consider the power rule model as an acceleration model, $\mu_i = \alpha V_i^{-\beta}$, $\alpha > 0$, for all V_i within a specified range. Suppose an accelerated life test is conducted according to the randomization procedure above, yielding the data $\{V_i, n_i, r_i, \hat{\mu}_i\}$, $i=1,\ldots,k$. The values of α and β are to be estimated so that inferences can be made about the mean lifetime μ_0 (and, hence, about the lifetime distribution) at the normal use stress V_0. In order to obtain asymptotically independent estimators of α and β, Singpurwalla (1971) amended the power rule model slightly, without changing its basic character, to $\mu_i = \alpha(\dot{V}/V_i)^{\beta}$, where $\dot{V}$ is the weighted geometric mean of the V_i's given by $\dot{V} = \prod_{i=1}^{k} V_i^{R_i}$ with $R_i = r_i / \sum_{j=1}^{k} r_j$. The maximum likelihood estimators of α and β, $\hat{\alpha}$ and $\hat{\beta}$, respectively, are then the solutions to the equations

$$\hat{\alpha} = \frac{\sum_{i=1}^{k} r_i \hat{\mu}_i (V_i/\dot{V})^{\hat{\beta}}}{\sum_{i=1}^{k} r_i} \tag{4.1}$$

and

$$\sum_{i=1}^{k} r_i \hat{\mu}_i (V_i/\dot{V})^{\hat{\beta}} \log(V_i/\dot{V}) = 0. \tag{4.2}$$

Since equation (4.2) is nonlinear, the solution $\hat{\beta}$ must be found numerically. Singpurwalla (1971) reported that, using the Newton-Raphson method, for various sets of data generated by computer, the solutions converged in five to ten iterations. Then the numerical solution $\hat{\beta}$ gives a unique value $\hat{\alpha}$ from (4.1). Thus, μ_0 is estimated by substituting $\hat{\alpha}$ and $\hat{\beta}$ into the amended power rule model at the normal use stress V_0.

Singpurwalla (1971) also obtained asymptotic expressions for the variances and covariance of the estimators, given by

$$\mathrm{var}(\hat{\alpha}) = (\sum_{i=1}^{k} r_i(\log(V_i/\dot{V}))^2)^{-1},$$

$$\mathrm{var}(\hat{\beta}) = \alpha^2(\sum_{i=1}^{k} r_i)^{-1},$$

and $\mathrm{cov}(\hat{\alpha},\hat{\beta}) = 0$. Since the maximum likelihood estimators are asymptotically unbiased and distributed as a bivariate normal, $\hat{\alpha}$ and $\hat{\beta}$ are asymptotically independent. This was the reason for using the amended power rule model. These results can be used for asymptotic inferences from the accelerated test data.

Shaked (1978) has studied the accelerated life testing problem for a parametric class of life distributions with linear hazard rate functions. Suppose a device has a life distribution of the general form

$$F^0(t) = 1 - \exp[-\alpha G(t) - \beta H(t)],\ t \geq 0, \qquad (4.3)$$

where α and β are unknown parameters and G and H are known differentiable functions of t, not depending on α and β. Some well-known life distributions in the literature have this form (Bain, 1974, and Prairie and Ostle, 1961, for example). In particular, the exponential distribution is a special case of (4.3).

In accelerated life tests where the underlying life distribution is assumed to be exponential at each stress level, the (constant) hazard rate r_i at stress level V_i under the power rule model is

$$r_i = \alpha V_i^P,\ \alpha > 0,\ -\infty < P < \infty, \qquad (4.4)$$

for $i=1,\ldots,k$. According to the Arrhenius model,

$$r_i = \alpha \exp(-P/V_i),\ \alpha < 0,\ -\infty < P < \infty. \qquad (4.5)$$

Since V_i is usually at least as large as V_0, acceleration means that $r_i \geq r_0$, where r_0 is the hazard rate under normal use conditions. Shaked (1978) also introduced an acceleration model which generalizes both (4.4) and (4.5), applied it to the life distributions (4.3), and obtained the maximum likelihood estimators of the unknown parameters. More explicitly, if $W = q(V)$, where q is a known positive increasing function of $V > 0$, then the linear hazard rate function of the life distribution under stress level V_i,

$$F_{V_i}(t) = 1 - \exp\{-W_i^P\,[\alpha G(t) + \beta H(t)]\},\ t \geq 0, \tag{4.6}$$

is given by

$$r_i(t) = W_i^P\,[\alpha g(t) + \beta h(t)],\ t \geq 0, \tag{4.7}$$

where $g = G'$ and $h = H'$. Note that if $q(V) = V$, $G(t) = t$, and $\beta = 0$, then (4.7) reduces to (4.4), and if $q(V) = \exp(-V^{-1})$, $G(t) = t$, and $\beta = 0$, (4.7) reduces to (4.5).

Estimates of α, β, and P based on accelerated test data are required to extrapolate to the life distribution and hazard rate function under the normal use stress V_0. Shaked (1978) showed how to obtain the maximum likelihood estimators, $\hat{\alpha}, \hat{\beta}$, and $\hat{P}$. Computation of these estimators requires numerical solution of nonlinear equations, but no more than one equation with one unknown must be solved at a time. Then to get the maximum likelihood estimator of the lifetime distribution function under the normal use stress, $\hat{\alpha}, \hat{\beta}$, and $\hat{P}$ can be substituted into (4.6) with $i=0$. Expressions for the asymptotic variance-covariance matrix of $\hat{\alpha}$, $\hat{\beta}$, and $\hat{P}$ involve integrals which must be computed numerically, but the computations can be readily performed by any reasonable numerical integration method.

5. SOME NONPARAMETRIC METHODS

In some situations, it may not be desirable or possible to determine a reasonable parametric form for the lifetime distribution of items under test at either the normal use stress or the accelerated stresses. This is particularly true during the development of a new device. For such situations, inference procedures have been developed which are nonparametric, at least in the sense that no specification of a family of parametric life distributions is given. Many of these procedures assume some kind of acceleration model whose unknown parameters are estimated from the accelerated data, and an extrapolation to the normal use stress is performed from the acceleration model in a manner similar to that of Section 4. If the particular form of the acceleration model is not assumed, it is usually necessary to have at least a few observations from the normal use stress level.

In this section some of the nonparametric methods for inference from accelerated life tests will be described briefly. These will include both types of methods, those without a specific acceleration model and those which utilize a particular form for an acceleration model. First, some results of Barlow and Scheuer (1971) will be summarized.

Barlow and Scheuer (1971) assumed that data were available from both the normal use life distribution F_0 and one accelerated life distribution F_1, where the exact forms of F_0 and F_1 were unspecified. It was assumed that F_1 and F_0 were related by a time transformation, $F_1(t) = F_0[\alpha(t)]$, where $\alpha(t)$ was unknown and $\alpha(t) \geq t$. The assumptions about F_0 and F_1 were that they were increasing failure rate average (IFRA) distributions; that is, $-t^{-1}\ln(1-F_i(t))$, $i=0,1$, were nondecreasing in t. The problem considered by Barlow and Scheuer was as follows: Let $Y_1 \leq \ldots \leq Y_m$ be ordered observations from F_0 and let $X_1 \leq \ldots \leq X_n$ be ordered observations from the accelerated life distribution F_1. The samples may be censored. Let F_{0m} and F_{1n} denote the empirical distribution functions. Then obtain estimates $\hat{F}_0$ and $\hat{F}_1$ such that $\hat{F}_0$ and $\hat{F}_1$ are IFRA, $\hat{F}_1(x) \geq \hat{F}_0(x)$ for all x, and $\hat{F}_0$ and $\hat{F}_1$ are closest to F_{0m} and F_{1n}, respectively, in a least squares sense. The estimate $\hat{F}_0$ of F_0 is obtained using data from both the accelerated and unaccelerated stress levels.

First, Barlow and Scheuer (1971) gave an estimator of a distribution function F which was assumed to be IFRA. To define their estimator, denote the k ordered distinct failure times out of n items on test with life distribution F by $X_1 \leq \ldots \leq X_k$ $(k \leq n)$, and let n_i denote the number of items surviving just before time X_i. The product-limit estimate of $\bar{F} = 1-F$ can be written as

$$\bar{F}_n(t) = \begin{cases} 1, & 0 \leq t < X_1 \\ \prod_{j=1}^{i} \left(\frac{n_j-1}{n_j}\right), & X_i \leq t < X_{i+1},\ i=1,\ldots,q-1 \\ 0, & t \geq X_q, \end{cases}$$

where $q=k$ if $k<n$ and $q=n$ if $k=n$ (no censoring), and $\bar{F}_n(t)$ is undefined for $t \geq X_q$ if $k<n$. Let $\lambda(t) = -t^{-1} \ln \bar{F}(t)$ and obtain an estimate of λ by a nondecreasing function $\hat{\lambda}_n$. Let $\lambda_n(X_i) = -X_i^{-1} \ln \bar{F}_n(X_i)$ and define

$$\hat{\lambda}_n(X_i) = \max_{s\leq i} \min_{i\leq t} \frac{\sum_{j=s}^{t} \lambda_n(X_j) F_n^*(X_j)}{\sum_{j=s}^{t} F_n^*(X_j)},$$

where $F_n^*(X_j) = \bar{F}_n(X_j^-) - \bar{F}_n(X_j^+)$. Then define

$$\hat{\lambda}_n(t) = \begin{cases} 0, & t < X_1 \\ \hat{\lambda}_n(X_i), & X_i \leq t < X_{i+1},\ i=1,\ldots,q-1 \\ +\infty, & t \geq X_q \end{cases}$$

so that the IFRA estimate of $\bar{F}$ is given by

$$\hat{\bar{F}}_n(t) = \begin{cases} 1, & t < X_1 \\ \exp[-\hat{\lambda}_n(X_i)t], & X_i \leq t < X_{i+1}, i=1,\ldots,q-1 \\ 0, & t \geq X_q. \end{cases}$$

Now, to estimate F_0 and F_1, suppose $X_1 \leq \ldots \leq X_k$ $(k \leq n)$ and $Y_1 \leq \ldots \leq Y_r (r \leq m)$ are (censored) samples from F_1 and F_0, respectively. Assume that (i) F_0 and F_1 are IFRA and (ii) $F_1(t) \geq F_0(t)$ for all $t \geq 0$. Let $\lambda_n(X_i) = -X_i^{-1} \ln \bar{F}_{1n}(X_i)$ and $\gamma_m(Y_j) = -Y_j^{-1} \ln \bar{F}_{0m}(Y_j)$, where $\bar{F}_{1n}$ and $\bar{F}_{0m}$ denote the product-limit estimators. Then find $\lambda_1, \ldots, \lambda_k$ and $\gamma_1, \ldots, \gamma_r$ to solve:

$$\text{minimize} \quad \{ \sum_{i=1}^{k} [\lambda_i - \lambda_n(X_i)]^2 F_{1n}^*(X_i)$$

$$+ \sum_{j=1}^{r} [\gamma_j - \gamma_m(Y_j)]^2 F_{0m}^*(Y_j)\}$$

$$\text{subject to} \quad 0 \leq \lambda_1 \leq \ldots \leq \lambda_k, \ 0 \leq \gamma_1 \leq \ldots \leq \gamma_r,$$

$$\text{and} \quad \hat{\bar{F}}_{0m}(t) \geq \hat{\bar{F}}_{1n}(t).$$

This problem can be solved by quadratic programming, and the solution is the best under the constraints in a least squares sense.

Barlow and Scheuer's (1971) estimators were obtained by monotonic regression techniques. Several other authors have considered nonparametric inference procedures for accelerated testing by assuming that the life distributions under the normal use stress and under an accelerated stress differed only by a scale parameter. Spurrier (1983) used a generalized Wilcoxon-Mann-Whitney test to compare two independent life tests subject to Type II censoring. Padgett and Wei (1982) estimated the ratio of scale parameters of two life distributions for arbitrarily right-censored data using a minimum distance procedure. Wei and Gail (1983) obtained nonparametric point and interval estimates of the ratio of scale parameters from arbitrarily right-censored samples using rank test statistics. McNichols and Padgett (1984) used the scale-change estimator of Padgett and Wei (1982) to obtain a nonparametric estimate of the life distribution F_{V_0} at normal use stress V_0 based on arbitrarily right-censored samples from k accelerated stress levels $V_1, \ldots, V_k$ as well as from the normal use condition.

As mentioned in Section 3, Shaked, Zimmer, and Ball (1979) assumed a general functional relationship between the life distributions of any two different stress levels as an acceleration model. No specific form for the life distributions was specified, however. The power rule model, Arrhenius model, Eyring model, and Chernoff's (1962) model were shown to be special cases of this acceleration model. Specifically, for every pair of stress levels V_i, V_j, $i \neq j$, $i,j = 0,1,\ldots,k$, it was assumed that a known function m existed so that the lifetime distributions satisfied $F_{V_j}(t) = F_{V_i}(m(\alpha, V_j, V_i, t))$, $t \geq 0$, where α was an unknown parameter (or vector of parameters). Let $X_{i\ell}^0$, $\ell=1,\ldots,n_i$, $i=1,\ldots,k$ be observations under stress levels $V_i, i=1,\ldots,k$, in the notation of Section 2. Hence, no observations are required from stress V_0. Here k, V_i's, and n_i's are selected and fixed in advance, with no censoring of the data.

For simplicity, Shaked, Zimmer, and Ball (1979) took $m(\alpha, V_j, V_i, t) = [g(\alpha, V_j)/g(\alpha, V_i)]t$, where g was a known function which was positive at each V and nondecreasing in V. Their procedure was then to estimate α from the accelerated test data by $\hat{\alpha}$ and then, for every $t \geq 0$, to estimate the life distribution at the normal use stress, F_{V_0}, directly from the data and $\hat{\alpha}$. The estimate of F_{V_0} was then used to estimate the mean lifetime, the percentiles, and other quantities in the normal use environment. An estimate of α was obtained by first estimating the scale factor $\theta_{ij} = g(\alpha, V_j)/g(\alpha, V_i)$ between F_{V_j} and F_{V_i}, $i \neq j$. Any consistent estimator of θ_{ij} could be used, such as $\hat{\theta}_{ij} = \bar{X}_i^0 / \bar{X}_j^0$, where $\bar{X}_i^0 = n_i^{-1} \sum_{\ell=1}^{n_i} X_{i\ell}^0$. Then α was estimated by a weighted average $\hat{\alpha} = \sum_{i \neq j} w_{ij}\, \hat{\alpha}_{ij}$, where the w_{ij}'s were chosen depending on g and $\hat{\theta}_{ij}$, and $\hat{\alpha}_{ij}$ was obtained from solving $\hat{\theta}_{ij} = g(\hat{\alpha}_{ij}, V_j)/g(\hat{\alpha}_{ij}, V_i)$. Using $\hat{\alpha}$, the observations from the k accelerated tests were rescaled by

$\tilde{X}^0_{i\ell} = [g(\hat{\alpha},V_j)/g(\hat{\alpha},V_i)]\ X^0_{i\ell}$, $\ell=1,\ldots,n_i$, $i=1,\ldots,k$. Finally, the empirical distribution function $\hat{F}_{V_0}$ of these $N = \sum_{i=1}^{k} n_i$ rescaled observations was used as an estimate of F_{V_0}.

Basu and Ebrahimi (1982) extended the results of Shaked, Zimmer, and Ball (1979) and of Shaked and Singpurwalla (1982) to the case of randomly right-censored samples $(X_{i\ell},\Delta_{i\ell})$ at each accelerated stress V_i. The choice of the acceleration model m was a power rule model. That is, $m(\alpha,V_j,V_i,t) = [g(\alpha,V_j)/g(\alpha,V_i)]t = (V_j/V_i)^{\alpha}t$ was used. The estimator of the scale factor $\theta_{ij} = (V_j/V_i)^{\alpha}$ was that obtained for the two-sample problem from arbitrarily right-censored samples by Padgett and Wei (1982). The value of α was estimated in a manner suggested by Shaked and Singpurwalla (1982), and the estimate was used to rescale the censored observations from the accelerated tests by $\tilde{X}_{i\ell} = (V_i/V_0)^{\hat{\alpha}}\ X_{i\ell}$, $\ell=1,\ldots,n_i$, $i=1,\ldots,k$. Then the product-limit estimator of Kaplan and Meier (1958) obtained from the N rescaled observations was used as an estimator for F_{V_0}. Basu and Ebrahimi (1982) also extended the results of Shaked and Singpurwalla (1982) to the case of I-component series systems (or independent competing risks) at each accelerated stress level.

Shaked and Singpurwalla (1982) combined the techniques of Sethuraman and Singpurwalla (1982) and of Shaked, Zimmer, and Ball (1979) for complete samples from each accelerated stress. The rescaling technique using the estimated α from the power rule model for the scale change parameters described above was used to estimate F_{V_0}. Tests for F_{V_0} being a Weibull distribution or a lognormal distribution were given, and uniform confidence bounds for F_{V_0} were provided.

Other nonparametric approaches to accelerated life testing have been developed. For example, Proschan and Singpurwalla (1980) adopted a Bayesian point of view without assumptions about the form of the lifetime distribution and the acceleration model. Shaked and Singpurwalla (1983) considered accelerated life testing wherein the stress on an unfailed item can increase at a preassigned test time. Such tests are called *step-stress tests*. Their approach was nonparametric in the sense that no assumptions were made about the underlying lifetime distribution at each stress. Their model for step-stress testing was based on the ideas of shock models and wear processes. Louis (1981) assumed a scale change to be the only difference between two life distributions and used a log rank statistic to generate an estimate of the scale change and a confidence interval-based test to compare the life distributions.

There are many other results on inference from accelerated life tests, too numerous to summarize here. It is hoped that the introduction to the topic and the references given in this paper will convey the basic ideas of accelerated testing and point out some of the recent results that have been obtained.

REFERENCES

Allen, W. R. (1959). "Inferences from Tests with Continually Increasing Stress," *J. Operations Research Society of America*, 303-312.

Bain, L. J. (1974). "Analysis for the Linear Failure-Rate Life-Testing Distribution," *Technometrics, 16*, 551-559.

Barlow, R. E., and Scheuer, E. M. (1971). "Estimation from Accelerated Life Tests," *Technometrics, 13*, 145-159.

Basu, A. P. and Ebrahimi, N. (1982). "Nonparametric Accelerated Life Testing," *IEEE Trans. Reliability, R-31*, 432-435.

Carfagno, S. P., and Gibson, R. J. (1980). *A Review of Equipment Aging Theory and Technology*, NP-1558, Research Project 890-1, Franklin Research Center, Philadelphia, Pa.

Chernoff, H. (1962). "Optimal Accelerated Life Designs for Estimation," *Technometrics, 4*, 381-408.

DeGroot, M., and Goel, P. K. (1979). "Bayesian Estimation and Optimal Designs for Partially Accelerated Life Testing," *Naval Res. Logistic Quart., 26*, 223-235.

Epstein, B., and Sobel, M. (1954). "Some Theorems Relevant to Life Testing from an Exponential Distribution," *Ann. Math. Statist.*, *25*, 373-381.

Escobar, L. A., and Meeker, W. Q., Jr. (1983). "Planning Accelerated Life Tests with Type II Censored Data," Technical Report, Dept. of Statistics, Iowa State Univ.

Hahn, G., and Meeker, W. Q., Jr. (1978). "A Comparison of Accelerated Test Plans to Estimate Survival Probabilities at a Design Stress," *Technometrics*, *20*, 245-247.

Kalbfleisch, J. D., and Prentice, R. L. (1980). *The Statistical Analysis of Failure Time Data*, New York: John Wiley & Sons.

Kaplan, E. L., and Meier, P. (1958). "Nonparametric Estimation from Incomplete Observations," *J. Amer. Statist. Assoc.*, *53*, 457-481.

Little, R. E., and Jebe, E. H. (1974). *Statistical Design of Fatigue Experiments*, New York: Halsted Press.

Louis, T. A. (1981). "Nonparametric Analysis of an Accelerated Failure Time Model," *Biometrika*, *68*, 381-390.

Mann, N. R. (1972). "Design of Over-Stress Life-Test Experiments When Failure Times Have the Two-Parameter Weibull Distribution," *Technometrics*, *14*, 437-451.

Mann, N. R., Schafer, R. E., and Singpurwalla, N. D. (1974). *Methods for Statistical Analysis of Reliability and Life Data*, New York: John Wiley & Sons.

McNichols, D. T., and Padgett, W. J. (1984). "Nonparametric Estimation from Accelerated Life Tests with Random Censorship," in *Reliability Theory and Models. Stochastic Failure Models, Optimal Maintenance Policies, Life Testing, and Structures*, Mohamed Abdel-Hameed, E. Çinlar, and J. Quinn, Eds., Orlando: Academic Press.

Meeker, W. Q., Jr. (1983a). "A Bibliography on Accelerated Testing," Technical Report, Dept. of Statistics, Iowa State Univ.

Meeker, W. Q., Jr. (1983b). "A Review of the Statistical Aspects of Accelerated Life Testing," Technical Report, Dept. of Statistics, Iowa State Univ.

Meeker, W. Q., Jr. (1984). "A Comparison of Accelerated Life Test Plans for Weibull and Lognormal Distributions and Type I Censoring," *Technometrics* (to appear).

Meeker, W. Q., Jr., and Nelson, W. (1975). "Optimum Accelerated Life Tests for Weibull and Extreme Value Distributions," *IEEE Trans. Reliability*, *R-24*, 321-332.

Meinhold, R. J., and Singpurwalla, N. D. (1984), "A Kalman Filter Approach to Accelerated Life Testing-- A Preliminary Development," in *Reliability Theory and Models. Stochastic Failure Models, Optimal Maintenance Policies, Life Testing, and Structures*, Mohamed Abdel-Hameed, E. Çinlar, and J. Quinn, Eds. Orlando: Academic Press.

Nelson, W. (1972). "Graphical Analysis of Accelerated Life Test Data with the Inverse Power Law Model," *IEEE Trans. Reliability, R-21*, 2-11, (correction, 195).

Nelson, W. (1974), "A Survey of Methods for Planning and Analyzing Accelerated Tests," *IEEE Trans. Electrical Insulation, EI-9*, 12-18.

Nelson, W. (1975). "Analysis of Accelerated Life Test Data -- Least Squares Methods for the Inverse Power Law Model," *IEEE Trans. Reliability, R-24*, 103-107.

Nelson, W. (1978). "Faster Accelerated Life Testing by Step-Stress -- Models and Data Analysis," General Electric Co., CR&D TIS Report No. 78 CRD051, Schenectady, New York.

Nelson, W. (1980). "Accelerated Life Testing - Step Stress Models and Data Analyses," *IEEE Trans. Reliability, R-29*, 103-108.

Nelson, W., and Kielpinski, T. (1975). "Optimal Accelerated Life Tests for Normal and Lognormal," *IEEE Trans. Reliability, R-24*, 310-320.

Nelson, W., and Kielpinski, T. (1976). "Theory for Optimum Accelerated Life Tests for Normal and Lognormal Life Distributions," *Technometrics, 18*, 105-114.

Nelson, W. B., and Meeker, W. Q. (1978). "Theory for Optimum Accelerated Censored Life Tests for Weibull and Extreme Value Distributions," *Technometrics, 20*, 171-177.

Padgett, W. J., and Wei, L. J. (1982). "Estimation of the Ratio of Scale Parameters in the Two-Sample Problem with Arbitrary Right-Censorship," *Biometrika, 69*, 252-256.

Prairie, R. R., and Ostle, B. (1961). "An Analysis of Some Relay Failure Data from Composite Exponential Population," *Technometrics, 3*, 423-428.

Proschan, F., and Singpurwalla, N. D. (1980). "A New Approach to Inference from Accelerated Life Tests," *IEEE Trans.Reliability, R-29*, 98-102.

Sethuraman, J., And Singpurwalla, N. D. (1982). "Testing of Hypotheses for Distributions in Accelerated Life Tests," *J. Amer. Statist. Assoc., 77*, 204-208.

Shaked, M. (1978). "Accelerated Life Testing for a Class of Linear Hazard Rate Type Distributions," *Technometrics, 20*, 457-466.

Shaked, M., and Singpurwalla, N. D. (1982). "Nonparametric Estimation and Goodness-of-Fit Testing of Hypotheses for Distributions in Accelerated Life Testing," *IEEE Trans. Reliability, R-31*, 69-74.

Shaked, M., and Singpurwalla, N. D. (1983). "Inference for Step-Stress Accelerated Life Tests," *J. Statist. Planning Inference, 7*, 295-306.

Shaked, M., Zimmer, W. J., and Ball, C. A. (1979). "A Nonparametric Approach to Accelerated Life Testing," *J. Amer. Statist. Assoc., 74*, 694-699.

Singpurwalla, N. D. (1971). "A Problem in Accelerated Life Testing," *J. Amer. Statist. Assoc., 66*, 841-845.

Spurrier, J. D. (1983). "Comparison of Two Independent Life Tests Subject to Type II Censoring," *Commun. Statist. - Theor. Meth., 12*, 1813-1829.

Wei, L. J., and Gail, M. H. (1983). "Nonparametric Estimation for a Scale-Change with Censored Observations," *J. Amer. Statist. Assoc., 78*, 382-388.

SMALL SAMPLE STUDIES FOR LINEAR DECREASING MEAN RESIDUAL LIFE

Ina Parks S. Howell

Department of Mathematical Sciences
Florida International University
Miami, Florida

The linear decreasing mean residual life (LDMRL) failure model has been proposed as an alternative to the popularly used failure models. This paper presents several small sample studies involing the LDMRL model. The small sample powers of the Hollander-Proschan test statistic (for constant mean residual life versus decreasing mean residual life) and the Gnedenko test statistic (for constant failure rate versus increasing failure rate) are compared for the alternative LDMRL distributions. We use Monte Carlo simulation to compare the two tests for sample sizes $n = 10, 15, 20, 30, 40, 50, 60$ and for LDMRL shape parameters $\gamma = 1.5, 2.0, 2.5, 3.0, 3.5, 4.0$. Point estimation and confidence intervals for the shape parameter and the maximum life expectancy parameter of the LDMRL model are briefly reviewed. We present small sample investigations of these confidence intervals. These studies suggest the need for a test of hypothesis concerning the shape parameter's value. A test to meet this need is given along with its simulated power for small samples.

ISBN 0-12-041420-1

1. INTRODUCTION AND SUMMARY

The exponential, gamma, Weibull, log-normal, and modified extreme value families are popular alternatives for lifetime distribution models. Bryson (1974) notes that an experimenter with reference to a long history of use of one particular family typically makes an a priori selection of the family and then estimates its parameters. Howell and Higgins (1983) propose the linear decreasing mean residual life (LDMRL) failure model as an alternative to the frequently used models. It has useful properties which are not well described by these other families (see Howell and Higgins (1983) for details). Note that nonparametric tests for decreasing mean residual life can be used to determine if there is evidence the failure times are from a DMRL family prior to estimating the parameters.

Let T represent the time-to-failure of some system (or component). The mean residual life function, m(t), is the expected remaining life of the system given that it has survived to time t. More formally,

$$m(t) = E[T-t \mid T>t].$$

The LDMRL model arises when m(t) is considered a linear decreasing function of time. A useful parametization of m(t) is

$$m(t) = \begin{cases} (T_o-t)/\gamma, & 0 \leq t \leq T_o,\ \gamma > 1,\ T_o > 0 \\ 0, & t > T_o. \end{cases}$$

In this form T_o and γ can be interpreted as the maximum life expectancy and the reciprocal rate of deterioration, respectively. The density function and failure rate function, respectively, are

$$f(t) = \begin{cases} ((\gamma-1)/T_o)(1-t/T_o)^{\gamma-2}, & 0 \leq t \leq T_o \\ 0, & t > T_o \end{cases}$$

and

$$r(t) = (\gamma-1)/(T_o - t),\ 0 \leq t < T_o.$$

Graphs of f(t) for $T_o = 1$ and selected values of γ are given in Figures 1 and 2.

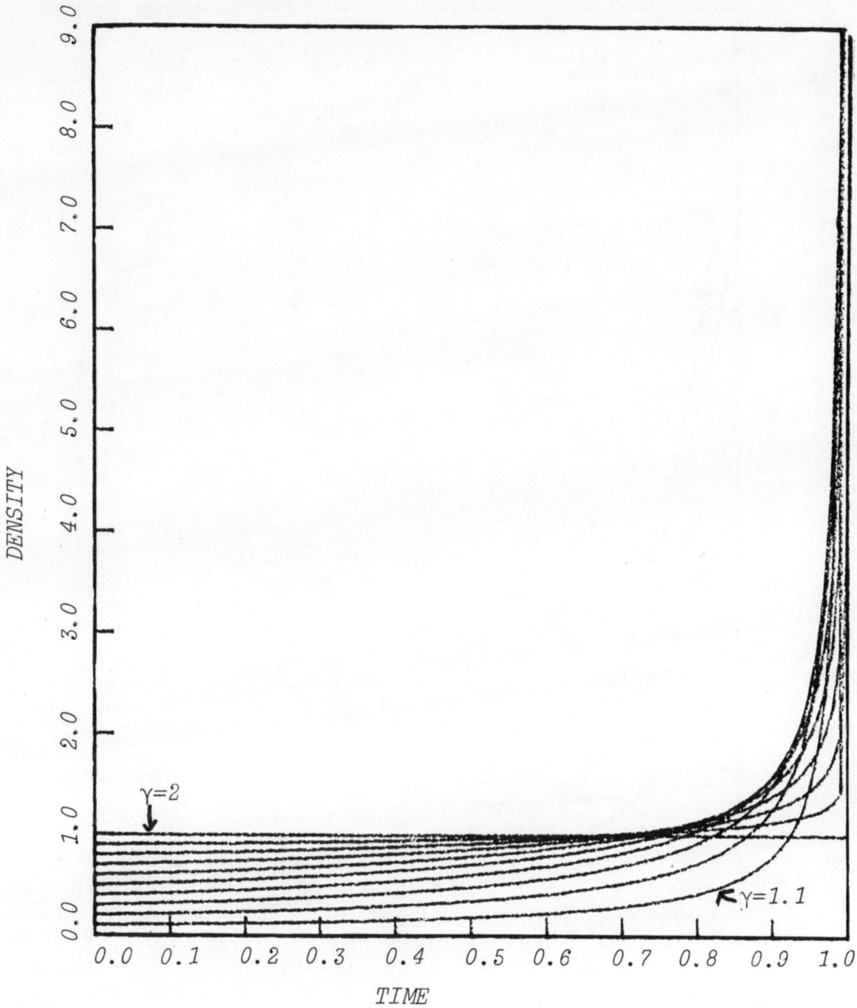

FIGURE 1. *The linear decreasing mean residual life probability density function for values of $\gamma \leq 2$ and $T_o = 1.0$. The vertical intercept is γ-1 for each function.*

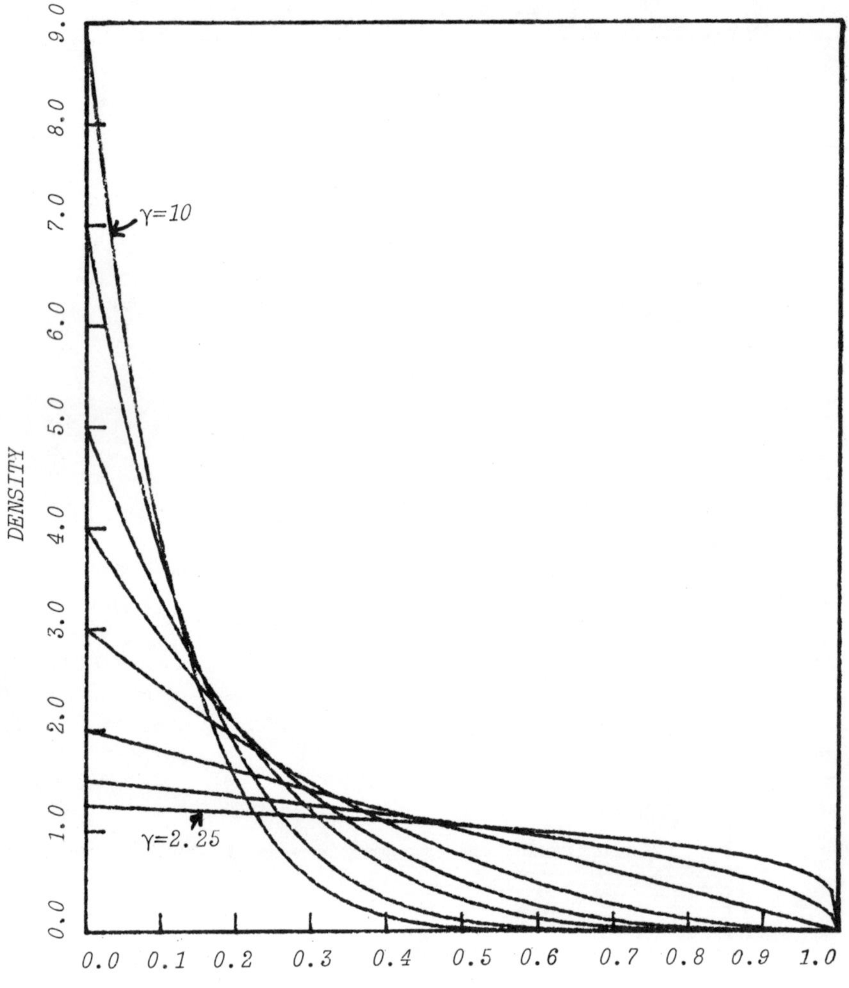

FIGURE 2. The linear decreasing mean residual life probability density function for values of $\gamma > 2$ and $T_o = 1.0$. The vertical intercept is γ-1 for each function.

Figure 1 (for $\gamma \leq 2$) represents lifetime data with an initial period in which relatively few failures occur followed by a short period in which many and eventually all items fail. Figure 2 (for $\gamma > 2$) represents lifetime data with many early failures with the fewer remaining failures occurring over a long period of time. If γ and T_o increase without bound in such a way that γ/T_o is a constant λ, then the LDMRL model approaches the exponential model $f(t) = \lambda e^{-\lambda t}$, $t > 0$. Our experience trying to fit the LDMRL model to data seems to indicate that for γ greater than 5 the corresponding LDMRL model and the exponential model with $\lambda = \gamma/T_o$ are indistinguishable for practical purposes, irrespective of the value of T_o.

Except for large values of γ, the characteristics of the LDMRL model do not resemble those of other failure models such as the exponential, Weibull, log-normal or gamma. Hence, when the hypothesis that the failure model is exponential is rejected in favor of the alternative that the model has decreasing mean residual life by the test of Hollander and Proschan (1975), the LDMRL provides a useful parametric model to consider. It supplies the user with an additional, yet simple tool for model building.

Simulation studies of the Hollander-Proschan (1975) test statistic and the Gnedenko, Belyayev, and Solovyev (1969) test statistic (commonly referred to as the Gnedenko test statistic) under the LDMRL alternative are presented in section 2. Point estimation and confidence intervals for the two parameters of the LDMRL are reviewed in section 3. The effect of small sample sizes on the confidence intervals is studied by Monte Carlo simulation in section 4. From this study the need for a test statistic for the LDMRL shape parameter is suggested. In section 5 we present a test statistic to meet this need. Its small sample power is investigated. Finally, section 6 illustrates the techniques on leukemia survival data.

TABLE 1. Simulated Powers of the Test Statistics

POWER AT $\gamma = 1.5$

	$\alpha = .01$		$\alpha = .05$		$\alpha = .10$	
N	H-P	G	H-P	G	H-P	G
10	.677	.797	.917	.864	.966	.916
15	.867	.875	.978	.922	.992	.958
20	.964	.970	.998	.979	1.000	.991
30	1.000	.997	1.000	.998	1.000	.999
40	1.000	.998	1.000	1.000	1.000	1.000
50	1.000	1.000	1.000	1.000	1.000	1.000
60	1.000	1.000	1.000	1.000	1.000	1.000

POWER AT $\gamma = 2$

	$\alpha = .01$		$\alpha = .05$		$\alpha = .10$	
N	H-P	G	H-P	G	H-P	G
10	.224	.492	.537	.611	.713	.723
15	.408	.588	.738	.695	.876	.802
20	.607	.725	.873	.812	.944	.875
30	.859	.851	.973	.914	.988	.953
40	.966	.928	.998	.962	1.000	.986
50	.988	.967	1.000	.984	1.000	.992
60	.997	.992	1.000	.997	1.000	.999

POWER AT $\gamma = 2.5$

	$\alpha = .01$		$\alpha = .05$		$\alpha = .10$	
N	H-P	G	H-P	G	H-P	G
10	.097	.325	.303	.445	.465	.585
15	.174	.384	.478	.516	.646	.655
20	.290	.487	.616	.629	.789	.762
30	.512	.638	.816	.749	.912	.846
40	.736	.764	.925	.852	.966	.914
50	.856	.813	.972	.889	.992	.943
60	.921	.935	.993	.953	.996	.975

TABLE 1. Simulated Powers of the Test Statistics

POWER AT $\gamma = 3.0$

	$\alpha = .01$		$\alpha = .05$		$\alpha = .10$	
N	H-P	G	H-P	G	H-P	G
10	.062	.295	.223	.421	.356	.566
15	.095	.322	.330	.429	.490	.566
20	.149	.391	.429	.504	.623	.653
30	.286	.468	.636	.593	.788	.720
40	.447	.595	.776	.710	.893	.825
50	.594	.649	.887	.764	.941	.865
60	.716	.798	.925	.868	.982	.928

POWER AT $\gamma = 3.5$

	$\alpha = .01$		$\alpha = .05$		$\alpha = .10$	
N	H-P	G	H-P	G	H-P	G
10	.033	.233	.170	.354	.286	.487
15	.070	.273	.235	.367	.402	.524
20	.097	.300	.303	.435	.472	.589
30	.150	.351	.448	.471	.623	.639
40	.259	.433	.566	.563	.744	.692
50	.385	.498	.705	.646	.823	.770
60	.496	.678	.779	.679	.897	.847

POWER AT $\gamma = 4.0$

	$\alpha = .01$		$\alpha = .05$		$\alpha = .10$	
N	H-P	G	H-P	G	H-P	G
10	.035	.216	.131	.303	.244	.433
15	.048	.223	.176	.325	.307	.469
20	.065	.241	.229	.370	.394	.525
30	.107	.301	.336	.401	.515	.541
40	.169	.344	.452	.478	.630	.619
50	.274	.409	.580	.546	.742	.673
60	.328	.561	.666	.658	.799	.774

2. POWERS OF THE HOLLANDER-PROSCHAN AND THE GNEDENKO TEST STATISTICS FOR THE LDMRL MODEL

Recall increasing failure rate (IFR) implies decreasing mean residual life (DMRL). The Gnedenko test statistic and the Hollander-Proschan test statistic can both be used to test the null hypothesis of constant MRL versus the alternative hypothesis DMRL. The powers of these two tests are compared against the one-sided alternative that the failure times follow a LDMRL distribution. The Gnedenko test is chosen for comparison to the Hollander-Proschan test in this study because Fercho and Ringer (1972) study four tests of contant failure rate versus non-constant failure rate and report highest power for Gnedenko's test. A Monte Carlo program on a Univac 1100/81A computer is used to study the effect of varying the sample size and shape parameter of the LDMRL model on the power of these two test statistics. Six values of γ(1.5, 2.0, 2.5, 3.0, 3.5, 4.0) and seven different sample sizes (10, 15, 20, 30, 40, 50, 60) are used. For each combination of γ and n, 1000 samples are done. The estimated powers are given in Table 1.

The test statistic, Q, given by Gnedenko, Belyayev, and Solovyev (1969) is an F criterion test with grouped failure times. The failure times are divided into two groups, one consisting of the first r_1 failure times and the other consisting of the remaining $r_2 = n - r_1$ failure times with r_1 being chosen arbitrarily. Gnedenko's test statistic is based on the normalized spacings given by

$$S_i = (n-i+1)(t_i-t_{i-1}),\ t_0 = 0,\ i = 1, 2, \ldots, n$$

and is

$$Q(r_1, r_2) = \sum_{i=1}^{r_1} S_i/r_1 \Big/ \sum_{i=r_1+1}^{n} S_i/r_2.$$

Under the null hypothesis $Q(r_1, r_2)$ is distributed as F with $2r_1$ and $2r_2$ degrees of freedom. If $Q(r_1, r_2) > F_{\alpha,2r_1,2r_2}$, the null hypothesis is rejected; and it is concluded that the failure rate is increasing. Hence the MRL is decreasing. In the simulation program r_1 and r_2 are the same as those used by Fercho and Ringer (1972), with r_1 equal to the greatest integer less than or equal to $n/2$ and $r_2 = n - r_1$.

The test statistic, V^*, given by Hollander and Proschan (1975) is

$$V^* = V/\bar{t}$$

where

$$V = n^{-4} \sum_{i=1}^{n} c_{in} t_i$$

and

$$c_{in} = \frac{4}{3} i^3 - 4ni^2 + 3n^2 i - \frac{1}{2}(n^3 - n^2 + i^2 - \frac{1}{3}i).$$

Langenberg and Srinivasan (1979) note that V^* can be expressed as a function of the normalized spacings S_i as

$$V^* = \sum_{i=1}^{n} e_{in} S_i \Big/ \sum_{i=1}^{n} S_i$$

with

$$e_{in} = \frac{1}{3}\left(\frac{i}{n}\right)^3 - \left(1 + \frac{1}{2n}\right)\left(\frac{i}{n}\right)^2 + \left(\frac{1}{2} + \frac{1}{2n} + \frac{1}{6n^2}\right)\frac{i}{n}$$

and give the exact null distribution of the test scientific $(210n)^{\frac{1}{2}}V^*$. Under the null hypothesis the S_i's are independent χ^2_2 variates. (Hollander and Proschan (1980) note a correction to the asymptotic variance of $n^{\frac{1}{2}}V^*$ when H_0 is not true. This does not affect the equations we use for our simulation program.)

For $n \leq 30$ the Gnedenko test statistic has greater power than the Hollander-Proschan test statistic in 52 of 72 combinations of γ and n. For $\gamma \geq 3$, it has greater power for 44 of 63 combinations of γ and n. Hence, the Gnedenko test is preferred for small n and large γ. For larger samples and γ small the Hollander-Proschan test holds its own, although the Gnedenko test is often rather close. The Gnedenko test is suggested for use against the alternative hypothesis that the distribution is LDMRL. The LDMRL distributions are not only in the DMRL class, but also the IFR class. From this observation, it is natural that the Gnedenko test competes well against the Hollander-Proschan test under LDMRL alternatives.

3. ESTIMATION OF PARAMETERS OF THE LDMRL MODEL

Let $t_1 < t_2 < t_3 < \ldots < t_n$ be the ordered observations in a random sample of size n from the LDMRL model. Various methods of estimating T_0 and γ based on the complete sample are proposed and compared by Howell (1981), and key results are summarized by Howell and Higgins (1983). They propose a simple two-step procedure based on the sample mean and the sample maximum.

This two-step procedure defined below generally has smaller mean squared errors than the other methods of estimation considered in their study. The initial estimate of T_0 is $\tilde{T}_0 = t_n$, the maximum observation. Since $E(T) = T_0/\gamma$ or $\gamma = T_0/E(T)$, the initial estimate of γ is $\tilde{\gamma} = \tilde{T}_0/\bar{t}$, where $\bar{t}$ is the sample mean. Since

$$E(t_n) = T_0 \left(1-\prod_{i=1}^{n} i(\gamma-1)/(1+i(\gamma-1))\right),$$

an adjustment for the bias of t_n can be made. Thus, the second step estimate of T_0 is

$$\hat{T}_0 = t_n \left(1-\prod_{i=1}^{n} i(\tilde{\gamma}-1)/(1+i(\tilde{\gamma}-1))\right)^{-1}.$$

The second step estimate of γ is

$$\hat{\gamma} = \hat{T}_0/\bar{t}.$$

It is possible to obtain approximate large sample confidence intervals for T_0 and γ. The distribution function of $y=t_n/T_0$ is given by

$$F_n(y) = (1-(1-y)^{\gamma-1})^n.$$

If γ is known, it is possible to find values of y_L and y_U such that $F_n(y_L) = \alpha/2$ and $F_n(y_U) = 1 - \alpha/2$. Thus, a $100 \times (1-\alpha)\%$ confidence interval for T_o is found by solving the inequality $y_L < t_n/T_o < y_U$ for T_o. For γ unknown, an approximate $100 \times (1-\alpha)\%$ confidence interval for T_o is found by substituting $\hat{\gamma}$ for γ in the confidence interval obtained when γ is known. This procedure gives the confidence interval

$$t_n/y_U < T_o < t_n/y_L$$

where

$$y_U = 1 - (1-(1 - \alpha/2)^{1/n})^{1/(\hat{\gamma} - 1)}$$

and

$$y_L = 1 -(1 - (\alpha/2)^{1/n})^{1/(\hat{\gamma}-1)}$$

The two-step estimate of $1/\gamma$ is $\bar{t}/\hat{T}_o$. Since $\hat{T}_o$ is a consistent estimate of T_o, the asymptotic distribution of $\bar{t}/\hat{T}_o$ is the same as that of $\bar{t}/T_o$ (the appropriate standardization are implicilly understood). By the central limit theorem, $\bar{t}$ has an approximate normal distribution with mean $1/\gamma$ and variance $(\gamma-1)/[\gamma^2(\gamma+1)n]$. Thus an approximate $100 \times (1-\alpha)\%$ confidence interval for $1/\gamma$ is

$$1/\hat{\gamma} + z_{\alpha/2}\sqrt{(\hat{\gamma}-1)/(\hat{\gamma}^2(\hat{\gamma}+1)n)}$$

where $z_{\alpha/2}$ is the upper $\alpha/2$ quantile for the standard normal distribution. Invert these limits to obtain a confidence interval for γ.

To illustrate the estimation procedures a sample of size 30 was generated from the LDMRL model with $\gamma=1.5$ and $T_o = 100$. The data is: 0.7, 5.6, 7.5, 20.3, 36.3, 40.4, 41.1, 42.9, 47.1, 49.7, 51.6, 55.9, 63.2, 69.8, 71.7, 73.5, 74.0, 80.0, 80.5, 89.3, 89.3, 93.1, 93.4, 96.1, 96.9, 97.1, 97.2, 98.1, 99.4, 99.9.

The one-step and two-step estimates are:

$$\tilde{T}_o = 99.9,\ \bar{t} = 65.4,\ \tilde{\gamma} = 1.528,$$

$$\hat{T}_o = 100.2,\ \hat{\gamma} = 1.53$$

For 95% confidence limits, $y_U = 1$ AND $y_L = .983$; so, (99.9, 101.6) is the 95% confidence interval for T_o. The 95% confidence interval for γ is (1.31, 1.83).

4. SMALL SAMPLE STUDY OF THE CONFIDENCE INTERVALS

Using the same values of γ and n as in the power study of section 3 and 1000 samples for each combination of γ and n, we find the proportion of time, the true value of the parameter was in the computed 95% confidence interval. For each simulation 3 different confidence intervals are obtained; the 95% confidence interval for T_o using the true value of γ, the 95% confidence interval for T_o using $\hat{\gamma}$, and the 95% confidence interval for γ using $\hat{\gamma}$. The results showed that the confidence interval for T_o using known γ is a 95% confidence interval at the .05 level of significance; however, when $\hat{\gamma}$ is used in place of the true γ to obtain a confidence interval for T_o, the simulated confidence levels were significantly smaller than 95% for all combinations of γ and n studied. Similarly the simulated confidence levels for estimating the true γ were significantly less than 95% for all values of γ except $\gamma=1.5$. For $\gamma=1.5$ samples of size 30 or greater had confidence levels of 95% at the .05 level of significance.

Multiple linear regression was used to analyze the simulated proportions as a function of γ and n for the approximate confidence intervals for T_o. The regression equation is

$$y = 1.02 + .0013n - .101\gamma$$

with $r^2=.974$ and 39 degrees of freedom. Both n and γ are found to be significant; therefore, for a fixed n if γ is increased by one unit the proportion in the confidence interval is decreased by .101. For a fixed value of γ an increase of one unit in sample size increases the proportion by only .0013. Thus to offset the effect of an increase of one unit in γ, the sample size would have to be increased by approximately 100.

For the proportion in the approximate confidence intervals for γ, the regression equation is

$$y = 1.12 + .0001n - .113\gamma$$

with $r^2=.949$ and 39 degrees of freedom. In this case, the sample size is not significant if the value of γ is in the model.

These results indicate the need to find a better estimator of γ when γ is larger. Hence the two-step procedure is modified with

$$\tilde{T}_o^* = (n + 1)t_n/n$$

and

$$\tilde{\gamma}^* = \tilde{T}_o^*/\bar{t}$$

used in place of $\tilde{T}_o$ and $\tilde{\gamma}$ to obtain $\hat{T}_o^*$ and $\hat{\gamma}^*$. For $\gamma=2$ $E(t_n)=nT_o/(n+1)$, hence $\tilde{T}_o^*$ is an unbiased estimator of T_o in this special case. Previous simulation studies (see Howell (1981)) indicate that for $\gamma>2$, the estimates of T_o and γ using $\hat{T}_o^*$ and $\hat{\gamma}^*$ are greater on the average than those using $\hat{T}_o$ and $\hat{\gamma}$. Both these procedures tend to underestimate the parameters for $\gamma>2$.

The simulation program was run again with the same samples using the initial estimate of $\tilde{T}_o^*$ and $\tilde{\gamma}^*$. For $\gamma\geq 3$ the proportion of time the true γ was in the simulated confidence interval increased significantly using $\hat{\gamma}^*$. The proportion in the simulated confidence intervals of T_o using $\hat{\gamma}^*$ was significantly better for $\gamma\geq 2.5$. The proportion in each of the five simulated confidence intervals for each combination of γ and n is given in Table 2.

TABLE 2. Percent in Confidence Interval

	γ	T_0 GIVEN γ	T_0 USING $\hat{\gamma}$	γ USING $\hat{\gamma}$	T_0 USING $\hat{\gamma}^*$	γ USING $\hat{\gamma}^*$
N=10	1.5	.950	.873	.918	.834	.870
	2.0	.947	.839	.904	.820	.850
	2.5	.950	.782	.853	.788	.860
	3.0	.961	.728	.807	.771	.873
	3.5	.954	.649	.743	.701	.823
	4.0	.948	.613	.690	.677	.792
N=15	1.5	.949	.875	.918	.850	.867
	2.0	.948	.858	.898	.837	.870
	2.5	.953	.772	.862	.783	.847
	3.0	.946	.730	.792	.763	.845
	3.5	.951	.656	.705	.707	.791
	4.0	.952	.639	.684	.679	.746
N=20	1.5	.949	.876	.920	.883	.893
	2.0	.945	.863	.902	.848	.871
	2.5	.953	.819	.856	.826	.860
	3.0	.949	.742	.825	.769	.855
	3.5	.949	.683	.741	.703	.792
	4.0	.952	.642	.644	.677	.729
N=30	1.5	.950	.917	.940	.905	.932
	2.0	.943	.872	.898	.875	.885
	2.5	.954	.821	.868	.833	.866
	3.0	.950	.739	.786	.763	.817
	3.5	.945	.709	.737	.730	.784
	4.0	.948	.652	.669	.680	.727
N=40	1.5	.939	.901	.944	.911	.947
	2.0	.940	.870	.911	.882	.908
	2.5	.934	.827	.873	.837	.865
	3.0	.941	.778	.822	.792	.838
	3.5	.946	.706	.735	.727	.773
	4.0	.937	.660	.657	.680	.687
N=50	1.5	.942	.905	.946	.909	.928
	2.0	.948	.894	.920	.902	.897
	2.5	.947	.848	.881	.852	.881
	3.0	.953	.782	.811	.802	.828
	3.5	.955	.719	.720	.738	.744
	4.0	.947	.680	.633	.699	.664
N=60	1.5	.953	.910	.953	.915	.947
	2.0	.951	.892	.930	.890	.925
	2.5	.942	.836	.883	.851	.882
	3.0	.947	.798	.812	.807	.844
	3.5	.948	.733	.710	.744	.729
	4.0	.943	.676	.638	.703	.668

Multiple linear regression was used to analyze the difference between the proportions using $\hat{\gamma}$ and $\hat{\gamma}^*$ as a function of γ and n for the appropriate confidence intervals for T_0 and γ. The sample size is not significant if γ is in the model for either parameter. The 95% confidence intervals for the mean difference, between the proportions is computed for each value of γ used in the study. These results are summarized in Table 3.

These results suggest that $\hat{\gamma}^*$ instead of $\hat{\gamma}$ be used to estimate γ whenever $\gamma \geq 3$. When $\gamma < 2$, $\hat{\gamma}$ is to be preferred. Therefore, we need to be able to decide whether to use $\hat{\gamma}$ or $\hat{\gamma}^*$. Since the shape of the distribution is different for $\gamma < 2$ and $\gamma > 2$, a test for $\gamma = 2$ versus $\gamma \neq 2$ is presented in the next section.

TABLE 3. Regression Results for Differences in the Proportions[a]

	T_0	γ
$y(\gamma,n)$	$32.9 - 18.4\gamma + .107n$	$76.4 - 37.0\gamma + .299n$
r^2	.551	.790
df	39	39
$y(\gamma)$	$36.4 - 18.4\gamma$	$86.0 - 37.0\gamma$
r^2	.554	.691
df	40	40
$\mu_{y/1.5}$	8.8 ± 7.92	30.5 ± 11.59
$\mu_{y/2.0}$	$-.4 \pm 5.95$	12.0 ± 8.70
$\mu_{y/2.5}$	-9.6 ± 4.65	-6.5 ± 6.81
$\mu_{y/3.0}$	-18.8 ± 4.65	-25.0 ± 6.81
$\mu_{y/3.5}$	-28.0 ± 5.95	-43.5 ± 8.70
$\mu_{y/4.0}$	-37.2 ± 7.92	-62.0 ± 11.59

[a]*Values are given as number per 100*

5. HYPOTHESIS TEST FOR THE LDMRL SHAPE PARAMETER

Since the large sample distribution of $\bar{t}/\hat{T}_o$ is the same as that of $\bar{t}/T_o$, and $\bar{t}$ is approximately normal with mean $1/\gamma$ and variance $(\gamma-1)/[\gamma^2(\gamma+1)n]$;

$$z = (1/\hat{\gamma}-1/\gamma)/\sqrt{\gamma^2(\gamma+1)n/(\gamma-1)}$$

is approximately standard normal. When $\gamma=2$, $z = (1/\hat{\gamma}-1/2)\sqrt{12n}$; therefore, to test the null hypothesis $\gamma=2$ versus the alternative hypothesis $\gamma\neq2$ the statistic z can be used as the test statistic for large n. Thus, if $z>z_{\alpha/2}$ we reject the null hypothesis and conclude that γ is less than 2; and if $z<-z_{\alpha/2}$, we reject the null hypothesis and conclude that γ is greater than 2.

Using the same combinations of γ and n as the previous studies and 1000 samples for each combination of γ and n. With $z_{\alpha/2}=1.96$, the proportions in the upper and lower tails as well as the acceptance region is computed. These values are given in Table 4. The simulation indicates that using critical values ±1.96 for all sample sizes studied α is at most .09. The proportions in the upper and lower tails are not significantly different.

Using $\pm t(.025, n-1)$ (where $t(.025, n-1)$ is the upper .025 quantile of a t-distribution with n-1 degrees of freedom) as the critical values for each sample size n studied, the simulation indicates that for all n studied α is at most .07. The proportions in the upper and lower tails are not significantly different. On the average the difference between the proportion using t and the proportion using z for the acceptance region is significantly larger than zero. Hence, using the critical values $\pm t(.025, n-1)$ rather than ±1.96 decreases α. Therefore for small n, the test statistic should be treated as a t-distribution.

TABLE 4. Percent in Regions for Hypothesis Test Of H_o: $\gamma = 2$ vs. H_A: $\gamma \neq 2$

		Using $z = \pm 1.96$			Using $t = \pm t(.025, n-1)$		
	γ	Lower	Middle	Upper	Lower	Middle	Upper
N=10	1.5	.002	.513	.485	.000	.630	.370
	2.0	.034	.927	.039	.016	.968	.016
	2.5	.127	.862	.011	.069	.927	.004
	3.0	.210	.784	.006	.129	.869	.002
	3.3	.308	.687	.005	.219	.778	.003
	4.0	.379	.619	.002	.292	.707	.001
N=15	1.5	.000	.397	.603	.000	.472	.528
	2.0	.041	.910	.049	.033	.940	.027
	2.5	.223	.770	.007	.165	.829	.006
	3.0	.373	.626	.001	.305	.694	.001
	3.5	.490	.509	.001	.424	.575	.001
	4.0	.612	.388	.000	.559	.441	.000
N=20	1.5	.000	.272	.728	.000	.317	.683
	2.0	.040	.915	.045	.035	.933	.032
	2.5	.250	.747	.003	.218	.780	.002
	3.0	.507	.493	.000	.457	.543	.000
	3.5	.666	.334	.000	.622	.378	.000
	4.0	.745	.255	.000	.716	.284	.000
N=30	1.5	.000	.122	.878	.000	.135	.865
	2.0	.042	.912	.046	.039	.923	.038
	2.5	.411	.589	.000	.381	.619	.000
	3.0	.706	.294	.000	.675	.325	.000
	3.5	.870	.130	.000	.855	.145	.000
	4.0	.915	.085	.000	.906	.094	.000
N=40	1.5	.000	.041	.959	.000	.046	.954
	2.0	.039	.924	.037	.035	.935	.030
	2.5	.503	.496	.001	.489	.510	.001
	3.0	.847	.153	.000	.839	.161	.000
	3.5	.938	.062	.000	.934	.066	.000
	4.0	.975	.025	.000	.974	.026	.000
N=50	1.5	.000	.022	.978	.000	.027	.973
	2.0	.045	.927	.028	.042	.931	.027
	2.5	.609	.391	.000	.592	.408	.000
	3.0	.918	.082	.000	.915	.085	.000
	3.5	.978	.022	.000	.976	.024	.000
	4.0	.988	.012	.000	.985	.015	.000
N=60	1.5	.000	.007	.993	.000	.008	.992
	2.0	.028	.931	.041	.025	.938	.037
	2.5	.687	.313	.000	.670	.330	.000
	3.0	.965	.035	.000	.963	.037	.000
	3.5	.991	.009	.000	.991	.009	.000
	4.0	.999	.001	.000	.999	.001	.000

6. EXAMPLE

Bryson and Siddiqui (1969) analyze a set of data from the National Cancer Institute consisting of the survival times in days from diagnosis to death of 43 patients suffering from chronic granulocytic leukemia. The survival times and corresponding nonparametric MRL estimates are given in Table 5. Based on the graphs of the nonparametric MRL and average failure rate estimates, they conclude that the MRL graph indicates DMRL while the sample average failure rate graph does not indicate the decreasing property as clearly. Hollander and Proschan (1975) use their test statistic on this data. They find a P-value of .08. Hall and Wellner (1981) infer that this data indicates a linear tail from 1000 to 2500 days with slope - .05, which is consistent with a uniform distribution on this interval.

TABLE 5. Survival Times and Nonparametric[1] Mean Residual Lifetimes

t_i	$\hat{m}(t_i)$	t_i	$\hat{m}(t_i)$	t_i	$\hat{m}(t_i)$
0	925.12	495	775.64	1109	745.38
7	939.98	497	802.30	1314	585.42
47	921.93	532	796.81	1334	616.82
58	933.70	571	788.12	1367	642.20
74	941.23	579	812.62	1534	528.00
177	860.29	581	845.87	1712	393.75
232	827.05	650	812.18	1784	367.71
273	807.89	702	796.38	1877	320.50
285	818.63	715	822.55	1886	373.80
317	809.76	779	798.47	2045	268.50
429	718.91	881	735.17	2056	343.33
440	730.03	900	758.29	2260	209.00
445	748.42	930	773.81	2429	80.00
455	763.03	968	784.87	2509	0.00
468	775.90	1077	724.14		

[1]*Data in this table is reproduced with the permission of the American Statistical Association.*

The two-step estimates for this data are $\hat{T}_0 = 2781.0$ and $\hat{\gamma} = 3.01$. We use these estimates to perform a Chi-square test for the null hypothesis that the distribution is LDMRL versus the alternative that the distribution is not LDMRL. We find a P-value > .10, hence the null hypothesis of LDMRL is not rejected. The 95% confidence interval for T_0 and γ are (2584.3, 3609.8) and (2.56, 3.95). The z-value for testing the null hypothesis that $\gamma = 2$ versus the alternative $\gamma \neq 2$ is -3.81 (P-value = .00007); hence, we find sufficient evidence to strongly suggest $\gamma > 2$. The Q statistic for this data using $r_1 = 21$ and $r_2 = 22$ is $Q(42,44) = 1.425803$ with p-value = .124.

REFERENCES

Bryson, M. C., "*Heavy-tailed distributions: properties and tests.*" *Technometrics 16*, 61-68 (1974).

Bryson, M. C. and Siddiqui, M. M. "*Some criteria for aging.*" *J. Am. Stat. Assoc. 64*, 1472-1483 (1969).

Fercho, W. W. and Ringer, L. J., "*Small sample power of some tests of the constant failure rate.*" *Technometrics 11*, 713-724 (1972).

Gnedenko, B. V., Belyayev, Yu. K. and Solovyev, A. D., *Math. Meth. Rel. Th.* Translated from the Russian by Scripta Technica. New York: Academic Press (1969).

Hall, W. J. and Wellner, J. A., Mean residual life, in *Statistics and Related Topics*, (M. Csorgo, D. A. Dawson, J. N. K. Rao and A. K. Md. E. Salih, ed.), 169-184 Amsterdam: North-Holland Publishing Company, (1981).

Howell, I. P. S., "*Parameter Estimation: Linear Decreasing Mean Residual Life Failure Model.* Unpublished Ph.D. Thesis, University of South Florida, Tampa, Florida, (1981).

Howell, I. P. S. and Higgins, J. J., "*Linear Decreasing Mean Residual Life Failure Model*" Submitted, (1983).

Hollander, M. and Proschan, F., "*Tests for the Mean Residual Life.*" *Biometrika 62*, 585-593 (1975).

Hollander, M. and Proschan, F., Correction to "*Tests for Mean Residual Life*" *62*, 585-593, *Biometrika 67*, 259 (1980).

Langenberg, P. and Srinivasan, R., *"Null distribution of the Hollander-Proschan statistic for decreasing mean residual life"*, *Biometrika 66*, 679-680 (1979).

Reliability Structures, Computability, and Approximations

RELIABILITY CALCULATION AND INDUCTIVE INFERENCE: AN OVERVIEW

R. E. Barlow

Operations Research Center
University of California-Berkeley
Berkeley, California

Reliability calculation and inductive inference are based on probability - its assessment and efficient calculation. In this survey, we overview recent developments regarding system reliability calculation, critique the two paradigms in inductive inference relative to reliability problems, and suggest promising directions for future research.

1. EFFICIENT CALCULATION OF SYSTEM RELIABILITY

In engineering reliability problems, a numerical estimate of system reliability is often required based on assessed component reliabilities and/or failure rates. Often the system reliability problem can be formulated in terms of a network graph or logic tree. The problem is then to calculate the probability that two or more vertices in the graph are connected or to calculate the probability that the TOP event in the logic tree occurs. In both cases, the complexity of the problem depends on the size and topology of the network or logic tree. For large systems, there is a need for efficient algorithms to solve these problems.

ISBN 0-12-041420-1

An algorithm is a *polynomial time algorithm* if for a problem of size n , its running time is bounded by a polynomial in n . Any algorithm that is not a polynomial time algorithm is generally referred to as an *exponential time algorithm.* In combinatorics, the so-called "satisfiability" problem is in the class NP-complete, i.e., given an arbitrary Boolean expression in product of sums form, determine whether or not there exists an assignment of values TRUE or FALSE to the variables which makes the entire expression TRUE (a Boolean expression can be obtained for every network in terms of Boolean indicators for the edges and nodes). A problem P is said to belong to the class *NP-complete* if (i) given a purported solution its validity can be checked in polynomial time, (ii) the existence of an algorithm to solve P in polynomial time implies the existence of algorithms to solve the satisfiability problem in polynomial time. It is generally believed that no polynomial time algorithm exists for any of the NP-complete problems. Any problem not NP-complete but which can be proved to be at least as hard as NP-complete problems is known as an *NP-hard problem.*

Most network and logic tree reliability problems are, in the worst case, NP-hard [Rosenthal (1975)], [Ball (1977 and 1980)]. Network reliability problems are, in a sense, more difficult than many standard combinatorial optimization problems. That is, given a tentative solution to a combinatorial problem, often its correctness can be determined in polynomial time. However, given a purported solution to a reliability problem, it cannot even be checked without computing the reliability of the system from the beginning.

Figure 1.1 is a computer communication network. The problem is to compute the probability that UC Santa Barbara (UCSB) can communicate with Carnegie Mellon University (CMU). Usually, it is assumed that transmission lines can fail but terminals are perfectly reliable. If terminals can fail but transmission lines are considered perfect, then a graph may be drawn where edges

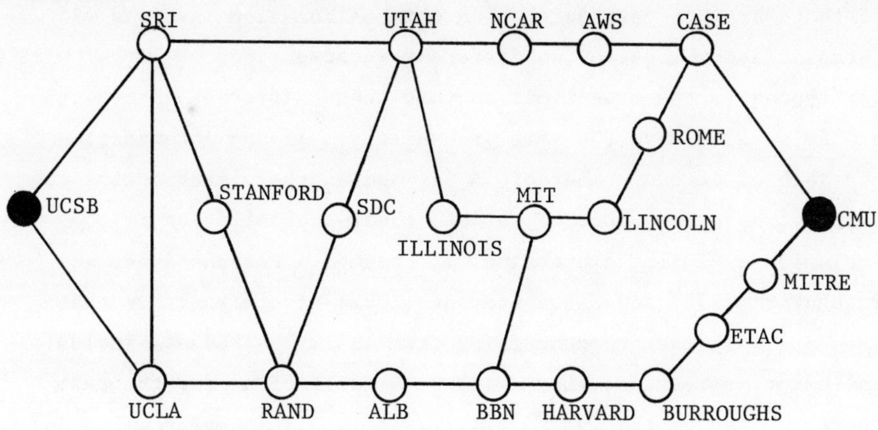

FIGURE 1.1. ARPA Computer Network.

represent terminals and nodes represent connections. In this example, the graph $G = (V,E)$, consists of vertices and edges. Edges are defined by their limiting vertices. The distinguished vertex set K for this problem consists of UCSB and CMU.

The reliability of a graph, G, with distinguished nodes K, is the probability, $R_K(G)$, that all elements of at least one minimal path-set are working or one minus the probability that all elements of at least one minimal cut-set have failed. Note that $R_K(G)$ depends on the distinguished node set $K \subseteq V$ as well as G. Different methods exist to evaluate this probability. These methods are quite general and can be used relative to any system reliability problem. We shall roughly classify them into three main classes. Let A_i denote the event that all elements in the ith minimal path-set are functional and $\bar{A}_i$ denote the complement of this event. Let p be the number of minimal path-sets.

(i) The Inclusion-Exclusion Method

$$R_K(G) = P\left[\bigcup_{i=1}^{p} A_i\right] = \sum_{i=1}^{p} P(A_i) - \sum_{i=1}^{n} \sum_{j<i} P[A_i A_j] + \cdots + (-1)^{p-1} P[A_1 A_2 \ldots A_p] . \tag{1.1}$$

If there are p path-sets, then this calculation involves $2^p - 1$ terms. In some cases, two different intersections of A_i's will correspond to the same event so that these different intersections of A_i's will have the same probability. If one intersection consists of an odd number of A_i's and another intersection consists of an even number of A_i's, their probabilities in the inclusion-exclusion expression will cancel. Satyanarayana and Prabhakar (1978) and Satyanarayana (1982) give algorithms that generate only the non-cancelling terms. In the *reduced* inclusion-exclusion expression, there will be a coefficient for the term corresponding to the event $A_1A_2 \ldots A_p$. (This event will, in general, correspond to several different intersections.) The coefficient for this term is called the *signed domination*, $d_K(G)$. This coefficient is the number of odd *formations* of G minus the number of even formations of G. (A *formation* is a set of minimal path-sets whose union constitutes the edge set E. It is odd (even) if it has an odd (even) number of minimal path sets in it.) The absolute value of this coefficient is called the *domination*, $D_K(G) = |d_K(G)|$. This number is a measure of the computational complexity of certain factoring algorithms for undirected networks.

This method is an exponential time method and chiefly of theoretical interest. It can be used to obtain bounds on $R_K(G)$. However, the time required to find the success sets may itself be excessive so that this method is not practical for most problems.

(ii) Sum of Disjoint Products Method

$$R_K(G) = P(A_1) + P(\bar{A}_1A_2) + \ldots + P(\bar{A}_1\bar{A}_2 \ldots \bar{A}_{p-1}A_p) . \quad (1.2)$$

For p path-sets, there are p terms but the time needed to generate each term may be exponential in p. Most methods based on Boolean techniques belong to this class. Such methods have been proposed by Agraham (1979), Fratta and Montanari (1973), Aggarwal, Misra and Gupta (1975), etc. One proposed by Lee (1979) for flow networks can easily be modified by using concepts of Satyanarayana (1982) to give the reliability of communication networks. This method, based on the idea of using backtracking and depth first search introduced by Gabow and Myers (1978), is the most efficient method in this class for the undirected all-terminal and the directed source-to-all-terminal problems; i.e., the distinguished set of vertices is the set of all vertices of the graph. The amount of computational work required is proportional to the number of spanning trees or rooted spanning arboresences. [See Deo (1974) for a detailed discussion of graph theory concepts.]

(iii) Pivotal Decomposition or Factoring

If $R_K(G \mid e)$ is the reliability of G under the condition that edge e is working and $R_K(G \mid \bar{e})$ the reliability of G under the condition that edge e is not working, then using the pivotal decomposition [cf. Barlow and Proschan (1981)],

$$R_K(G) = p_e R_K(G \mid e) + (1 - p_e) R_K(G \mid \bar{e}) \ . \qquad (1.3)$$

$R_K(G)$ for any graph G can be computed by repeated application of this decomposition. Undirected graphs have some special properties that can be used to simplify this method. If the vertices are assumed to be working, then $R_K(G \mid e)$ is the same as $R_K(G_e)$ where G_e is the graph obtained from G by deleting edge e and merging its end points. Similarly, $R_K(G \mid \bar{e})$ is the same as $R_K(G - e)$ where $G - e$ is the graph with e deleted but no vertex deleted. These schemes have been discussed by a number of writers, among them, Moskowitz (1958) and Misra (1970). In the literature, the above relationship has been referred to as the *factoring theorem*. It is important to note that this method can be employed using the graph representation but *without* knowing the minimal path-sets. However, unless some kind of probability reductions are performed (e.g. parallel and series reductions) after each pivot, this method will be equivalent to state space enumeration; i.e., in this case, it will be no better than the previous two methods. Using the optimal pivoting strategy discussed in Satyanarayana and Chang (1983), the computational complexity is proportional to the *domination*, $D_K(G)$, defined in the discussion of the inclusion-exclusion method. Although this factoring method is, in the worst case, also an exponential time method, it will usually be a great improvement over the two previous methods. However, the method does require that edges fail independently of each other.

A theoretical breakthrough occurred in 1978 with the publication by Satyanarayana and Prabhakar which first introduced the idea of domination into network reliability. This paper was concerned with directed graphs and suggested that directed cyclic graphs have domination zero, a result rigorously proved by R. R. Willie (1980). In 1981, Satyanarayana and Chang (see their 1983 paper) first noticed the connection between domination theory and the computational complexity of factoring algorithms. This was followed by the notion of minimal domination and the development of linear time algorithms for special structure graphs [cf. R. K. Wood (1982)].

The probability measure associated with the graphs in the literature surveyed is extremely simplistic and unrealistic for many practical network reliability problems. However, a theoretical basis now exists for more research into realistic probability measures. At the present time, there is no comparable theoretical basis for analyzing the computational complexity of logic (or fault) trees. These structures are perhaps more useful in practice than networks. Further research is badly needed in this field.

2. THE TWO PARADIGMS IN INDUCTIVE INFERENCE

When it comes to the problem of how to use statistical data in the assessment of probability, we find fundamental disagreement in the literature and the profession. At the heart of this disagreement is the nature and meaning of probability itself. In one of his editorials, Ralph Evans (1982) hit on a source of the disagreement. It concerns the use and interpretation of the phrase *random variable*. Is the life of a system a random variable or is it simply an unknown quantity - until observed? Is the average life of a population of systems a random variable or is it simply an unknown quantity which - were all lifetimes of the population measured - would become known? Probability is a measure of our uncertainty about a lifetime or an event - but does it *exist* in its own right - objectively, outside our minds? These questions are at the heart of a continuing controversy (but no longer a dialogue) among mathematical statisticians, both in universities and in industry. The controversy is between statisticians who call themselves "Bayesian" versus "classical" statisticians. (There are many practicing statisticians who do not understand the depth of the difference.) The controversy is both deep and, at times, bitter. In some sense, it has been going on since 1763! [T. Bayes (1763).]

Although it is not well understood, the *Bayesian Approach*, based on the concept of probability as a degree of belief, is a *way of thinking* about problems concerning uncertainty - it is *not* just another technique or method. Similarly, the classical approach based on the frequency interpretation of probability is a *way of thinking* about problems concerning uncertainty - it is not just another technique or method. The classical approach breaks down when little or no "frequency" data is at hand. Consequently, we have applied statisticians and engineers using the Bayesian approach when the data is dominated by engineering judgement - this has already happened in seismic hazard analysis, and in nuclear power reactor risk assessment. It has already happened in business management education (the Harvard School of Business) and in engineering economics (Stanford Department of Engineering Economics). Control theory engineers employing the Kalman filter are also using a technique which is fully Bayesian.

The controversy is over two fundamentally different "paradigms" or world views. There is a parallel between the Bayesian controversy and the controversy concerning quantum physics. In physics, there has been a longstanding (since 1900 or so) controversy concerning objectivity in the physical world [H. R. Pagels (1982)]. Does the electron "exist" or have a location apart from the observer? Quantum physics has proved, in a practical sense, to be useful in electronics and nuclear physics. It works at the microcosmic level - classical physics does not! Both classical statistics and classical physics "seem" to work at the macrocosmic level - actually there are many counterexamples contradicting even the logic of classical statistics at the large sample level. The Bayesian approach provides a methodology for learning and updating information - the classical statistics approach does not. In my opinion, this is very crucial. The analogy between quantum physics and the Bayesian approach is only a strong parallel. Although quantum physics talks about "probability waves," this concept would correspond more closely to what the Bayesian calls "chance" or "propensity."

Some of the best discussions of the Bayesian approach are Lindley [(1983), (1982), (1978) and (1971)]. In my view, the Bayesian approach should be taught in courses concerning reliability theory. Especially needed is support for research and development of software which will provide efficient methods for employing the Bayesian approach.

References

1. Abraham, J. A., *IEEE Trans. Reliability, R-28*, 58 (1979).
2. Aggarwal, K. K., Misra, K. B., and Gupta, J. S., *IEEE Trans. Reliability, R-24*, 83 (1975).
3. Ball, M. O., "Network Reliability Analysis: Algorithms and Complexity," Ph.D. Thesis, Dept. of Operations Research, Cornell University (1977).
4. Ball, M. O., *Networks, 10*, 153 (1980).
5. Barlow, R. E., and Proschan, F., "Statistical Theory of Reliability and Life Testing," To Begin With, Silver Springs, MD (1981).
6. Barlow, R. E., "Set Theoretic Signed Domination for Coherent Systems," ORC 82-1, Operations Research Center, University of California, Berkeley (1982).
7. Bayes, T., reprinted in *Biometrika, 45*, 293 (1958), with a note by G. A. Bernard (1763).
8. Deo, N., "Graph Theory with Applications to Engineering and Computer Science," Prentice-Hall, New Jersey (1974).
9. Evans, Ralph, *IEEE Trans. Reliability, R-31*, 321 (1982).
10. Fratta, L., and Montanari, U. G., *IEEE Trans. Circuit Theory, C-20*, 203 (1973).
11. Gabow, H. N., and Myers, E. W., *SIAM J. Computing, 7*, 280 (1978).
12. Lee, S. H., *IEEE Trans. Reliability, R-29*, 24 (1980).
13. Lindley, D. V., "Bayesian Statistics: A Review," SIAM, Philadelphia, PA (1971).
14. Lindley, D. V., *Scand. J. Statist. 5*, 1 (1978).
15. Lindley, D. V., *in* "Some Recent Advances in Statistics," (J. Tiago de Oliveira and Benjamin Epstein, eds.), Academic Press, London (1982).
16. Lindley, D. V., *J. Inst. Statist. 32*, 1 (1983).
17. Misra, K. B., *IEEE Trans. Reliability, R-19*, 146 (1970).
18. Moskowitz, F., *AIEE Trans. Commun. Electron, 77*, 627 (1958).
19. Pagels, H. R., "The Cosmic Code," Bantam, New York (1982).
20. Rosenthal, A., *in* "Reliability and Fault Tree Analysis," (R. E. Barlow, Fussell, J. B., and Singpurwalla, N. D., eds.), p. 133, SIAM, PA (1975).
21. Satyanarayana, A. and Prabhakar, A., *IEEE Trans. Reliability, R-27*, 82 (1978).
22. Satyanarayana, A., *IEEE Trans. Reliability, R-31*, 23 (1982).
23. Satyanarayana, A., and Chang, M. K., *Networks, 13*, 107 (1983).
24. Willie, R. R., *Networks, 10*, 71 (1980).
25. Wood, R. K., "Polygon-to-Chain Reductions and Extensions for Reliability Evaluations of Undirected Networks," Ph.D. Thesis, Operations Research Center, University of California, Berkeley (1982).

DECOMPOSITIONS FOR MULTISTATE MONOTONE SYSTEMS

H. W. Block[1]
T. H. Savits[1]

Department of Mathematics and Statistics
University of Pittsburgh
Pittsburgh, Pennsylvania

Multistate structure functions can be decomposed as sums of integrals of binary functions. Decompositions of this type are surveyed and applications are given.

1. INTRODUCTION

A fundamental notion in reliability is the concept of a structure function. Structure functions have been used to describe binary monotone systems and more recently to describe more complex systems. See El-Neweihi and Proschan (1983) for a review. Various representations for these functions have been given and often these representations have been used to derive bounds on the structure function. In this paper we give an up-to-date survey of results in this area.

In Section 2 we examine structure functions and their representations in the binary case. Section 3 considers similar re-

[1] *Supported by ONR Contract N00014-76-C-0839.*

ISBN 0-12-041420-1

sults for finite discrete structure functions with emphasis on the results contained in Block and Savits (1982a). Finally decompositions for structure functions with continuous states are given in Section 4. The proofs of these results are contained in Block and Savits (1982b).

In general we limit our discussion to $A \subset \mathbb{R}^n$. For $\underline{x} = (x_1,\ldots,x_n)$, $\underline{y} = (y_1,\ldots,y_n)$ in $\mathbb{R}^n$, $\underline{x} \le \underline{y}$ means $x_i \le y_i$ for $i = 1,\ldots,n$, $\underline{x} < \underline{y}$ means $x_i < y_i$ for $i = 1,\ldots,n$ and $\underline{x} \lneq \underline{y}$ means $\underline{x} \le \underline{y}$ and $\underline{x} \ne \underline{y}$. We also define

$$L_Q(\underline{x}) = \{\underline{y}|\underline{y} \le \underline{x}\},\ L_Q^o(\underline{x}) = \{\underline{y}|\underline{y} < \underline{x}\},$$

$$U_Q(\underline{x}) = \{\underline{y}|\underline{y} \ge \underline{x}\},\ U_Q^o(\underline{x}) = \{\underline{y}|\underline{y} > \underline{x}\}.$$

We also use the notation $\phi(j_i,\underline{x})$ to mean $\phi(x_1,\ldots,x_{i-1},j,x_{i+1},\ldots,x_n)$.

2. BINARY STRUCTURE FUNCTIONS

In general *binary monotone structure functions* are simply functions ϕ: $\{0,1\}^n \to \{0,1\}$ which are nondecreasing in each variable. These functions describe a system with n components where each component is in either of two modes, functioning (1) or failed (0). The system is only functioning or failed and depends only on which components are functioning or failed. The assumption that ϕ is nondecreasing reflects the assumption that improving a component (from failed to functioning) cannot cause a functioning system to fail. We assume the conditions $\phi(\underline{0}) = 0$ and $\phi(\underline{1}) = 1$. It is often assumed that each component is relevant, i.e., that each argument of the function plays some role.

For a binary monotone structure function with relevant components, the following representations hold:

$$\phi(\underline{x}) = \max_{1\le j\le p} \min_{i\in P_j} x_i = \min_{1\le j\le k} \max_{i\in K_j} x_i$$

where K_j are min cut sets and P_j are min path sets. See Barlow and Proschan (1975) for the definition of these quantities and a derivation of the result. This representation can be given in terms of min path vectors and min cut vectors. For example for min path vectors:

$$\phi(\underline{x}) = \max_{\underline{y}\in U_1} \min_{1\le i\le n} \alpha_{iy_i}(\underline{x})$$

where $U_1 = \{\underline{y}: \phi(\underline{y}) = 1, \underline{x} \lneq \underline{y} \Rightarrow \phi(\underline{x}) = 0\}$ and $\alpha_{iy_i}(\underline{x}) = \begin{cases} 1 \text{ if } y_i \le x_i \\ 0 \text{ if } y_i > x_i \end{cases}$. To see this note that $\phi(\underline{x}) = 1 \Leftrightarrow$ there is $\underline{y} \in U_1$ such that $\underline{x} \ge \underline{y}$. Thus $\{\underline{x}|\phi(\underline{x}) = 1\} = \bigcup_{\underline{y}\in U_1} \{\underline{x}|\underline{x} \ge \underline{y}\}$ and so the result follows.

The min cut vector representation is given by

$$\phi(\underline{x}) = \min_{\underline{y}\in L_0} \max_{1\le i\le n} \beta_{iy_i}(\underline{x})$$

where $L_0 = \{\underline{y}|\phi(\underline{y}) = 0, \underline{y} \lneq \underline{z} \Rightarrow \phi(\underline{z}) = 1\}$ and $\beta_{iy_i}(\underline{x}) = \begin{cases} 1 \text{ if } y_i < x_i, \\ 0 \text{ if } y_i \ge x_i. \end{cases}$

Both of the above representations facilitate obtaining bounds on $P\{\phi(\underline{X}) = 1\}$ where $\underline{X}$ is a random state vector. See Sections 2.3, 2.4 and 2.6 of Barlow and Proschan (1975).

3. DISCRETE MULTISTATE STRUCTURE FUNCTIONS

A *multistate monotone structure (MMS)* function is a function

$$\Phi: \{0,1,\ldots,M\}^n \to \{0,1,\ldots,M\}$$

which is nondecreasing in each variable and such that $\Phi(\underline{0}) = 0$, $\Phi(\underline{M}) = M$. A simple decomposition which is useful in carrying out inductive proofs is

$$\Phi(\underline{x}) = \sum_{j=0}^{M} \Phi(j_i,\underline{x})\, I_{\{j\}}(x_i) \quad \text{for } i = 1,\ldots,n,$$

where

$$I_{\{j\}}(x_i) = \begin{cases} 1 \text{ if } x_i = j, \\ 0 \text{ otherwise.} \end{cases}$$

See Lemma 3.18 of El-Neweihi and Proschan (1982). A decomposition in terms of binary valued functions was given by Block and Savits (1982a):

$$\Phi(\underline{x}) = \sum_{k=1}^{M} \phi_k(\underline{\alpha}(\underline{x}))$$

where

$$\phi_k(\underline{\alpha}(\underline{x})) = \max_{\underline{y} \in U_k} \min_{1 \le i \le n} \alpha_{iy_i}(\underline{x}),$$

$$\alpha_{iy_i}(\underline{x}) = \begin{cases} 1 \text{ if } y_i \le x_i, \\ 0 \text{ if } y_i > x_i, \end{cases}$$

and

$$U_k = \{\underline{y} : \Phi(\underline{y}) \ge k,\ \underline{z} \lneq \underline{y} \Rightarrow \Phi(\underline{z}) < k\}.$$

An alternative representation is

$$\Phi(\underline{x}) = \sum_{k=0}^{M-1} \psi_k(\underline{\beta}(\underline{x}))$$

where

$$\psi_k(\underline{\beta}(\underline{x})) = \min_{\underline{y} \in L_k} \max_{1 \le i \le n} \beta_{iy_i}(\underline{x}),$$

$$\beta_{iy_i}(\underline{x}) = \begin{cases} 1 \text{ if } y_i < x_i, \\ 0 \text{ if } y_i \ge x_i, \end{cases}$$

and

$$L_k = \{\underline{y} : \Phi(\underline{y}) \le k,\ \underline{z} \gneq \underline{y} \Rightarrow \Phi(\underline{z}) > k\}.$$

Here U_k is the set of critical upper vectors for level k and the first representation follows from the fact that $\Phi(\underline{x}) \ge k$ if and only if there is $\underline{x}^o \in U_k$ such that $\underline{x} \ge \underline{x}^o$. The alternate representation follows similarly or by duality arguments.

Both representations decompose the MMS Φ into a sum of binary functions ϕ_k or ψ_k. It is then possible to obtain bounds as in the binary case. Let $\underline{X} = (\underline{X}_1, \ldots, \underline{X}_n)$ be a random state vector. Let

$$P_i(j) = P\{X_i \leq j\},\ \bar{P}_i(j) = 1-P_i(j),\ i=1,\ldots,n,\ j=0,\ldots,M,$$

$$P(j) = P\{\Phi(\underline{X}) \leq j\},\ \bar{P}(j) = 1-P(j).$$

Theorem 1. Let Φ be an MMS and $k = 0,1,\ldots,M-1$.

(a) The following bounds always hold:

$$\max_{\underline{y}\in U_{k+1}} P\{\bigcap_{i=1}^{n} \{X_i > y_i-1\}\} \leq \bar{P}(k) \leq \min_{\underline{y}\in L_k} P\{\bigcup_{i=1}^{n} \{X_i > y_i\}\}.$$

(b) If the X_i are associated, then

$$\prod_{\underline{y}\in L_k} P\{\bigcup_{i=1}^{n} \{X_i > y_i\}\} \leq \bar{P}(k) \leq \coprod_{\underline{y}\in U_{k+1}} P\{\bigcap_{i=1}^{n} \{X_i > y_i-1\}\}.$$

(c) If the X_i are independent, then

$$\prod_{\underline{y}\in L_k} \coprod_{i=1}^{n} \bar{P}_i(y_i) \leq \bar{P}(k) \leq \coprod_{\underline{y}\in U_{k+1}} \prod_{i=1}^{n} \bar{P}_i(y_i-1).$$

Proof: See Block and Savits (1982).

Note: See also Butler (1982) and Natvig (1982) who obtained similar results independently.

4. CONTINUOUS MULTISTATE STRUCTURE FUNCTIONS

A function Φ: $[0,\infty)^n \to [0,\infty)$ which is nondecreasing and Borel measurable is called a *multistate monotone structure* (MMS) function. In the more general case (which follows after the R_+^n case) the domain is assumed to be a Borel measurable set $\Delta \subset R^n$. All proofs are contained in Block and Savits (1982b).

4.1 The R_+^n Case

Let Φ: $[0,\infty)^n \to [0,\infty)$ be right continuous and nondecreasing (i.e. a right continuous MMS). Let $U_t = \{\underline{x}|\Phi(\underline{x}) \geq t\}$, which is a closed upper set (i.e. $\underline{x}\in U_t$, $\underline{x} \leq \underline{y} \Rightarrow \underline{y}\in U_t$), and set $\phi(\underline{x},t) = I_{U_t}(\underline{x}) = I_{[0,\Phi(\underline{x})]}(t)$.

Theorem 2. $\Phi(\underline{x}) = \int_0^\infty \phi(\underline{x},t)dt.$

Results about Upper Sets: Let $T \subset \mathbb{R}$ and let A, $\{A(t) | t \in T\}$ be upper sets. Then

(i) $\bigcup_{t\in T} A(t)$, $\bigcap_{t\in T} A(t)$ are upper sets,

(ii) $A = \bigcup_{\underline{x}\in A} U_Q(\underline{x})$ where $U_Q(x) = \{\underline{y}: \underline{x} \leq \underline{y}\}$,

(iii) $\bar{A}$ and A^0 are upper sets (where $\bar{A}$ and A^0 are the closure and interior of A).

We need the following terminology. The *closed upper hull* of the set A is defined by $\cap\{C \supset A | C \text{ closed upper}\}$ (i.e., the smallest closed upper set containing A). We say $\underline{x}$ is a *lower extreme point* if $L_Q(\underline{x}) \cap \bar{A} = \{\underline{x}\}$ where $L_Q(\underline{x}) = \{\underline{y}: \underline{y} \leq \underline{x}\}$. Also let $E_L(A) = \{\text{lower extreme points of A}\}$.

Theorem 3. Let A be a closed upper set in R^n_+. Then

1) $A = \bigcup_{\underline{y}\in E_L(A)} U_Q(y)$.

2) A is the closed upper hull of $E_L(A)$.

Corollary. For a closed upper set $A \subset R^n_+$,

$$I_A(\underline{x}) = \max_{\underline{y}\in E_L(A)} \min_{1\leq i\leq n} \alpha_{iy_i}(\underline{x})$$

where $\alpha_{it}(\underline{x}) = 1$ if $x_i \geq t$ and 0 otherwise.

Remarks. 1) The above results can be extended to any closed upper set $A \subset R^n$ if A is *lower bounded*, i.e. ther is a $\underline{y} \in R^n$ such that $\underline{y} \leq \underline{x}$ for all $\underline{x} \in A$.

2) For an open upper set $A \subset R^n$ which is lower bounded

$$A = \bigcup_{\underline{y}\in E_L(A)} U^0_Q(\underline{y})$$

where $U^0_Q(\underline{y}) = \{\underline{x} | \underline{y} < \underline{x}\}$.

Theorem 4. Let A be a closed upper set in R^n_+, and assume there is an $E \subset R^n_+$ such that

$$I_A(\underline{x}) = \max_{\underline{y} \in E} \min_{1 \leq i \leq n} \alpha_{iy_i}(\underline{x}).$$

Then $E \supset E_L(A)$.

The results of Theorem 4 show that the representation in terms of the lower extreme points of A is minimal.

The preceding results are summarized in the following.

Min Path Representation

Let $\Phi\colon [0,\infty)^n \to [0,\infty)$ be nondecreasing and right continuous. Then

$$\Phi(\underline{x}) = \int_0^\infty \phi(\underline{x},t)dt, \text{ for all } \underline{x} \in R_+^n$$

where

$$\phi(\underline{x},t) = I_{U_t}(\underline{x}) = \max_{\underline{y} \in P_t} \min_{1 \leq i \leq n} \alpha_{iy_i}(\underline{x})$$

and $P_t = E_L(U_t)$.

Note. 1) Since $\underline{x} \in R_+^n$ we can replace $\min_{1 \leq i \leq n} \alpha_{iy_i}(\underline{x})$ by $\min_{i:y_i \neq 0} \alpha_{iy_i}(\underline{x})$ in the above.

2) Since $\underline{y} \in P_t \iff \Phi(\underline{y}) \geq t$ and $\Phi(\underline{z}) < t$ for all $\underline{z} \lneq \underline{y}$, the above is analogous to the min path result in the discrete case.

4.2 *Generalizations*

The set A where $A \subset \Delta \subset R^n$ is said to be *upper relative to* Δ if for $\underline{x} \in A$ and $\underline{y} \in \Delta$ such that $\underline{x} \leq \underline{y}$ implies $\underline{y} \in A$. We also say that A is upper in Δ. The set $B \subset \Delta$ is *lower relative to* Δ if $\Delta \setminus B$ is upper in Δ. It follows that A is upper relative to Δ if and only if there is an upper set U in R^n such that $A = U \cap \Delta$. Also let $\Phi\colon \Delta \to [0,\infty)$, $\Delta \subset R^n$ be an MMS (i.e. Φ Borel measurable and nondecreasing on Δ). Also let $U_t = \{\underline{x} \in \Delta \mid \Phi(\underline{x}) \geq t\}$ for all $t \geq 0$, $\phi(\underline{x},t) = I_{U_t}(\underline{x}) = I_{[0,\Phi(\underline{x})]}(t)$ for $t \geq 0$, $\underline{x} \in \Delta$, $U_t^0 = \{\underline{x} \in \Delta \mid \Phi(x) > t\}$, $t \geq 0$, and $\xi(\underline{x},t) = I_{U_t^0}(\underline{x}) = I_{[0,\Phi(\underline{x}))}(t)$, for $t \geq 0$, $\underline{x} \in \Delta$.

Simple Results

1) $\Phi(\underline{x}) = \int_0^\infty \phi(\underline{x},t)dt = \int_0^\infty \xi(\underline{x},t)dt$, for $t \geq 0$ and $\underline{x} \in \Delta$.

2) For upper sets A, $\{A(t), t \in T\}$ in Δ,

i) $\bigcup_{t\in T} A(t)$, $\bigcap_{t\in T} A(t)$ are upper in Δ,

ii) $A = \bigcup_{\underline{x}\in A} U_Q(\underline{x}) \cap \Delta$, and

iii) if Δ is a product set then the closure of A in Δ and the interior of A in Δ are upper in Δ.

For an upper set A in Δ, we define the set of *lower extreme points of A* relative to Δ by

$$E_L^{\Delta}(A) = \{\underline{y} \in \Delta \mid L_Q(\underline{y}) \cap \bar{A} \cap \Delta = \{\underline{y}\}\}.$$

Theorem 5. For closed $\Delta \subset \mathbb{R}^n$, closed upper A in Δ and A lower bounded,

$$A = \bigcup_{\underline{y}\in E_L^{\Delta}(A)} U_Q(\underline{y}) \cap \Delta$$

and

$$I_A(\underline{x}) = \max_{\underline{y}\in E_L^{\Delta}(A)} \min_{1\leq i\leq n} \alpha_{iy_i}(\underline{x}), \ \underline{x} \in \Delta$$

where $\alpha_{iy_i}(\underline{x}) = 1$ if $y_i \leq x_i$ and 0 otherwise.

Theorem 6. Let Δ be closed and lower bounded in R^n, $\Phi\colon \Delta \to [0,\infty)$ be an upper semicontinuous MMS. Then

$$\Phi(\underline{x}) = \int_0^{\infty} \phi(\underline{x},t)\,dt, \ \underline{x} \in \Delta$$

where

$$\phi(\underline{x},t) = I_{U_t}(\underline{x}) = \max_{\underline{y}\in P_t} \min_{1\leq i\leq n} \alpha_{iy_i}(\underline{x}), \ \underline{x} \in \Delta$$

and $P_t = E_L^{\Delta}(U_t)$.

For a lower set B in Δ we define the set of *upper extreme points of B relative to* Δ by

$$E_U^{\Delta}(B) = \{\underline{z} \mid U_Q(\underline{z}) \cap \bar{B} \cap \Delta = \{\underline{z}\}\}.$$

Theorem 7. For a closed $\Delta \subset R^n$, closed lower B in Δ and B upper bounded,

$$B = \bigcup_{\underline{z}\in E_U^{\Delta}(B)} L_Q(\underline{z}) \cap \Delta$$

and

$$I_B(\underline{x}) = \max_{\underline{z} \in E_U^\Delta(B)} \min_{1 \le i \le n} \beta_{iz_i}(\underline{x}), \quad \underline{x} \in \Delta$$

where $\beta_{iz_i}(\underline{x}) = 1$ if $x_i \le z_i$ and 0 otherwise.

Let $L_t = \{\underline{y} | \Phi(\underline{x}) \le t\}$, and $L_t^{(o)} = \{y | \Phi(\underline{x}) < t\}$.

Theorem 8. For finite $\Delta \subset R^n$ and $\Phi: \Delta \to [0,\infty)$ MMS

$$\phi(\underline{x},t) = I_{U_t}(\underline{x}) = \max_{\underline{y} \in P_t} \min_{1 \le i \le n} \alpha_{iy_i}(\underline{x})$$

$$= \min_{\underline{z} \in K_t^{(o)}} \max_{1 \le i \le n} \alpha_{iz_i}^{(o)}(\underline{x})$$

where

$$P = E_L^\Delta(U_t), K_t^{(o)} = E_U^\Delta(L_t^{(o)}) \text{ and } \alpha_{iz_i}^{(o)}(\underline{x}) = 1 \text{ if } x_i > z_i$$

and 0 otherwise.

Note. In general $L_t^{(o)}$ is not closed so Theorem 7 does not apply. However if Φ is continuous, L_t is closed.

Theorem 9. Let $\Phi: \Delta \to [0,\infty)$ be a continuous MMS on a compact Δ. Then for $t \ge 0$ and $\underline{x} \in \Delta$

$$\Phi(\underline{x},t) = \max_{\underline{y} \in P_t} \min_{1 \le i \le n} \alpha_{iy_i}(\underline{x}) \ge \min_{\underline{z} \in K_t} \max_{1 \le i \le n} \alpha_{iz_i}^{(o)}(\underline{x}) = \xi(\underline{x},t)$$

where P_t is as in Theorem 8 and $K_t = E_U^\Delta(L_t)$.

4.3 Bounds

Let $\underline{X} = (X_1,\ldots,X_n)$ be a random state vector (i.e. $\underline{X}$ takes values in Δ), $F_i(t) = P\{X_i \le t\}$, $\bar{F}_i(t) = 1 - F_i(t)$, Φ be a continuous MMS on compact Δ, $F(t) = P\{\Phi(X) \le t\}$, and $\bar{F}(t) = P\{\Phi(X) > t\}$ $\bar{F}(t-) = P\{\Phi(X) \ge t\})$.

Theorem 10. Let Φ be a continuous MMS on a compact Δ, and let $t \ge 0$. Then

$$\sup_{\underline{y} \in P_t} P(\bigcap_{i=1}^n \{X_i \ge y_i\}) \le \bar{F}(t-), \quad \bar{F}(t) \le \inf_{\underline{z} \in K_t} P\{\bigcup_{i=1}^n \{X_i > z_i\}\}.$$

Theorem 11. Let Φ be a continuous MMS on compact Δ (assumptions are also made on X, see Block and Savits (1982b)). Then for all but a countable number of $t \geq 0$, the following results hold.

a) For X_i associated

$$\prod_{\underline{z}\in K_t} P\{\bigcup_{i=1}^{n}\{X_i > z_i\}\} \leq \bar{F}(t-) = \bar{F}(t) \leq \coprod_{\underline{y}\in P_t} P\{\bigcap_{i=1}^{n}\{X_i \geq y_i\}\}$$

where Π, $\amalg$ are appropriately defined if K_t, P_t are infinite.

b) For X_i independent

$$\prod_{\underline{z}\in K_t} \coprod_{i=1}^{n} \bar{F}(z_i) \leq \bar{F}(t-) = \bar{F}(t) \leq \coprod_{\underline{y}\in P_t} \prod_{i=1}^{n} \bar{F}(y_i).$$

5. APPLICATIONS OF THE CONTINUOUS RESULTS

5.1 Barlow-Wu Structure Function

Consider a binary coherent system with min path sets $P_1, \ldots, P_p$ and min cut sets $K_1, \ldots, K_k$. Define

$$\zeta(\underline{x}) = \max_{1\leq r\leq p} \min_{i\in P_r} x_i = \min_{1\leq s\leq k} \max_{i\in K_p} x_i .$$

Then $\zeta\colon R_+^n \to [0,\infty)$ is a continuous multistate monotone structure function. This type of function has recently been studied by Baxter (1983).

Theorem 12. $\zeta(\underline{x}) = \int_0^\infty \zeta(\underline{x},t)\,dt$

where

$$\zeta(\underline{x},t) = I_{U_t}(\underline{x}) = \max_{\underline{y}\in E_L(U_t)} \min_{1\leq i\leq n} \alpha_{iy_i}(\underline{x}) = \zeta(\alpha_t(\underline{x}))$$

and

$$\alpha_t(\underline{x}) = (\alpha_{1t}(\underline{x}), \ldots, \alpha_{nt}(\underline{x})).$$

5.2 *Homogeneous Functions*

Theorem 13. Let Φ: $[0,\infty)^n \to [0,\infty)$ be right continuous nondecreasing and such that $\Phi(t\,\underline{x}) = t\,\Phi(\underline{x})$, for all $t \geq 0$. Then

$$\Phi(\underline{x}) = \max_{\underline{y} \in E_L(U_1)} \min_{y_i \neq 0} y_i^{-1} x_i$$

where

$$E_L(U_1) = \{\underline{y} \mid \Phi(y) \geq 1 \text{ and } \underline{x} \lneq \underline{y} \Rightarrow \Phi(\underline{x}) < 1\}.$$

REFERENCES

[1] Barlow, R.E. and Proschan, F. (1975). *Statistical Theory of Reliability and Life Testing*. Holt, Rinehart and Winston, New York.

[2] Barlow, R.E. and Wu, A.S. (1978). Coherent systems with multistate components. *Math of O.R. 3*, 275-281.

[3] Baxter, L.A. (1983). Continuous structures. Unpublished manuscript.

[4] Block, H.W. and Savits, T.H. (1982a). A decomposition for multistate monotone systems. *J. Appl. Probab. 19*, 391-402.

[5] Block, H.W. and Savits, T.H. (1982b). Continuous multistate structure functions. University of Pittsburgh Technical Report No. 82-27.

[6] Butler, D.A. (1982). Bounding the reliability of multistate systems. *Operations Research 30*, 355-374.

[7] El-Neweihi, E. and Proschan, F. (1982). Degradable systems: a survey of multistate system theory. FSU Statistics Report M637.

[8] Natvig, B. (1982). Two suggestions of how to define a multistate coherent system. *Adv. Appl. Probab. 14*, 434-455.

COMPUTING THE RELIABILITY OF K OUT OF N SYSTEMS

Philip J. Boland[1]

Department of Mathematics
University College Dublin
Belfield, Dublin 4, Ireland

Frank Proschan[1]

Department of Statistics
The Florida State University
Tallahassee, Florida

We survey some of the more important theoretical results about the structure of the reliability function of a k out of n system, and indicate how these results may be used to obtain easily calculable bounds for the reliability of a specified k out of n system.

[1]*This research was sponsored by the Air Force Office of Scientific Research Grant AFOSR 82-K-0007.*

ISBN 0-12-041420-1

I. INTRODUCTION

A system of n components which functions if and only if at least k of the components function is called a k out of n system. Parallel systems are 1 out of n systems, fail safe systems are n-1 out of n systems, and series systems are n out of n systems. We make the assumption that the n components in the system operate independently of one another. Therefore if $\underline{p} = (p_1,\ldots,p_n)$ is the vector of component reliabilities and $\underline{\epsilon} = (\epsilon_1,\ldots,\epsilon_n)$ represents any vector with components equal to zeroes or ones, then

$$h_k(\underline{p}) = \sum_{\underline{\epsilon},\epsilon_1+\ldots+\epsilon_n \geq k} p_1^{\epsilon_1} \cdots p_n^{\epsilon_n}(1-p_1)^{1-\epsilon_1} \cdots (1-p_n)^{1-\epsilon_n}$$

is the probability that k or more of the components function. We refer to $h_k(\underline{p}): [0,1]^n \to [0,1]$ as the reliability function of a k out of n system with independent components. For parallel, failsafe, and series systems, the reliability function $h_k(\underline{p})$ is a quite manageable function, even for large values of n. In general, however, the behavior of the function $h_k(\underline{p})$ on $[0,1]^n$ can be quite complex, and the calculation of $h_k(\underline{p})$ at a single vector of component reliabilities can be quite cumbersome. In the relatively simple case of a 5 out of 8 system for example, evaluating $h_5(\underline{p})$ would usually involve calculating 93 products of 8 numbers. In this expository paper we survey some of the important theoretical results which have been obtained about the reliability function $h_k(\underline{p})$, and indicate how these may be used to reduce calculations and obtain good bounds for the reliability of a k out of n system.

II. BASIC THEORETICAL RESULTS

Hoeffding (1956) considers the problem of finding the maximum and minimum of $h_k(p_1,\ldots,p_n)$ subject to the constraint that $\sum_{i=1}^{n} p_i$ is held fixed. His results are presented in terms of the number of successes in independent Bernoulli trials. It is clear, however, that if S is the number of successes in n independent Bernoulli trials with respective success probabilities $p_1,\ldots,p_n$, then $\mathrm{Prob}(S\geq k) = h_k(p_1,\ldots,p_n)$. For a given vector $\underline{p}=(p_1,\ldots,p_n)$ of component probabilities, we let

$$\overline{p} = \sum_{i=1}^{n} p_i/n, \qquad [\sum_{i=1}^{n} p_i]$$

be the greatest integer less than or equal to $\sum_{i=1}^{n} p_i$, and $(1,\ldots,1, \sum_{i=1}^{n} p_i - [\sum_{i=1}^{n} p_i], 0,\ldots,0)$ be the vector with $[\Sigma p_i]$ of its n coordinates equal to 1. Hoeffding proves in particular the following theorem:

Theorem 1. For any vector $\underline{p} = (p_1,\ldots,p_n)$ of component probabilities we have:

$$1 = h_k(1,\ldots,1, \sum_{i=1}^{n} p_i - [\sum_{i=1}^{n} p_i], 0,\ldots,0)$$

$$\geq h_k(p_1,\ldots,p_n) \geq h_k(\overline{p},\ldots,\overline{p})$$

$$\text{if} \quad \sum_{i=1}^{n} p_i \geq k ,$$

while

$$0 = h_k(1,\ldots,1, \sum_{i=1}^{n} p_i - [\sum_{i=1}^{n} p_i], 0,\ldots,0)$$

$$\leq h_k(p_1,\ldots,p_n) \leq h_k(\bar{p},\ldots,\bar{p})$$

$$\text{if} \quad \sum_{i=1}^{n} p_i \leq k - 1$$

Hoeffding also establishes upper and lower bounds for $h_k(\underline{p})$ for the case when $k-1 < \sum_{i=1}^{n} p_i < k$; however these bounds are considerably more complicated.

We present the first of four figures in an attempt to give some geometrical insight into the behavior of $h_k(\underline{p})$. For any α such that $0 \leq \alpha \leq n$, we define $\lambda_\alpha = \{\underline{p}: \sum_{i=1}^{n} p_i = \alpha$ and $0 \leq p_i \leq 1$ for $i=1,\ldots,n\}$. Although the convex set λ_α is the intersection of the hyperplane $\{\underline{x}: \sum_{i=1}^{n} x_i = \alpha\}$ with the n-dimensional unit cube, it will be represented goemetrically by a "line" in the following figures. The n-dimensional cube itself is represented by a "square", and by the "diagonal" we will mean the line of points all of whose coordinates are equal. Theorem 1 says that $h_k(\underline{p})$ restricted to λ_α assumes a maximum of one at the 'extremities' of λ_α and a minimum of $h_k(\alpha/n,\ldots,\alpha/n)$ along the diagonal whenever $\alpha \geq k$, while it assumes a maximum of $h_k(\alpha/n,\ldots,\alpha/n)$ on the diagonal and a minimum of 0 at the extremities of λ_α whenever $\alpha \leq k - 1$.

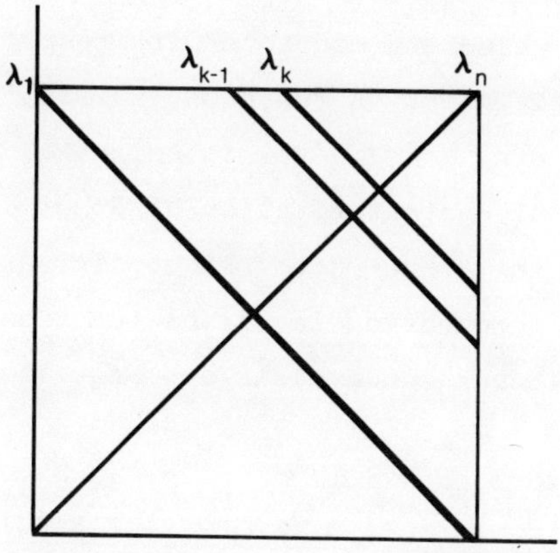

FIGURE 1

The theory of majorization and Schur functions provides an elegant format for presenting many of the results concerning the reliability function $h_k(\underline{p})$. Given a vector $\underline{x} = (x_1,\ldots,x_n)$, let $x_{[1]} \geq x_{[2]} \geq \ldots \geq x_{[n]}$ denote a decreasing rearrangement of $x_1,\ldots,x_n$: (we write $\underline{x} >^m \underline{y}$) if

$$\sum_{i=1}^{j} x_{[i]} \geq \sum_{i=1}^{j} y_{[i]} \quad \text{for } j=1,\ldots,n-1$$

and

$$\sum_{i=1}^{n} x_{[i]} = \sum_{i=1}^{n} y_{[i]} \quad .$$

Hardy, Littlewood and Polya (1952) show that $\underline{x} >^m \underline{y}$ if and only if there exists a doubly stochastic matrix Π such that $\underline{y} = \underline{x}\Pi$. If $\underline{x} >^m \underline{y}$, then the coordinates of $\underline{x}$ are more "dispersed" than those of $\underline{y}$ (See Marshall and Olkin (1979) for an excellent treatment of majorization.) Schur functions are real valued functions which are monotone with respect to the partial ordering of majorization. A function h with the property that $\underline{x} >^m \underline{y} \Rightarrow h(\underline{x}) \geq (\leq)\ h(\underline{y})$ is called *Schur convex (Schur concave)*. A useful characterization of Schur convexity (concavity) is given by the Schur-Ostrowski condition, which states that a differentiable permutation invariant function h defined on R^n is Schur convex (Schur concave) if and only if

$$(x_i - x_j)\left(\frac{\partial h}{\partial x_i} - \frac{\partial h}{\partial x_j}\right) \geq (\leq)\ 0 \quad \text{for all } i, j \text{ and } \underline{x} \in R^n .$$

Some useful comparisons for $h_k(\underline{p})$ can be obtained by working with the hazard transform. The hazard transform z_k: $[0,\infty)^n \to [0,\infty)$ of h_k is defined by

$$z_k(R_1,\ldots,R_n) = -\log h_k(e^{-R_1},\ldots,e^{-R_n}) .$$

Pledger and Proschan (1971) prove the following theorem about z_k:

Theorem 2. The hazard transform z_k of a k out of n system is increasing and Schur concave on $[0,\infty)^n$.

A particularly interesting corollary of Theorem 2 is that $h_k(p_1,\ldots,p_n) \geq h_k(p_G,\ldots,p_G)$ where p_G is the geometric mean $(\prod_{i=1}^{n} p_i)^{1/n}$. Together with Theorem 1 this implies that

$$h_k(\bar{p},\ldots,\bar{p}) \geq h_k(p_G,\ldots,p_G)$$

whenever

$$\sum_{i=1}^{n} p_i \le k-1 \quad .$$

Gleser (1975) uses majorization to obtain refinements of the inequalities established by Hoeffding. If S is the number of successes in n independent trials with respective success probabilities $p_1,\dots,p_n$, then Samuels (1965) shows that Prob(S=j) is unimodal with mode at $[\sum_{i=1}^{n} p_i]$ or $[\sum_{i=1}^{n} p_i+1]$. Using this fact together with the Schur-Ostrowski condition, Gleser proves the following theorem:

Theorem 3. The reliability function $h_k(\underline{p})$: $[0,1]^n \to [0,1]$ is Schur convex in the region where $\sum_{i=1}^{n} p_i \ge k+1$ and Schur concave in the region where $\sum_{i=1}^{n} p_i \le k-2$.

Figure 2 may help to illustrate Theorem 3. The region shaded by vertical lines indicates where $h_k(\underline{p})$ is Schur convex, while the area shaded by horizontal lines indicates where $h_k(\underline{p})$ is Schur concave.

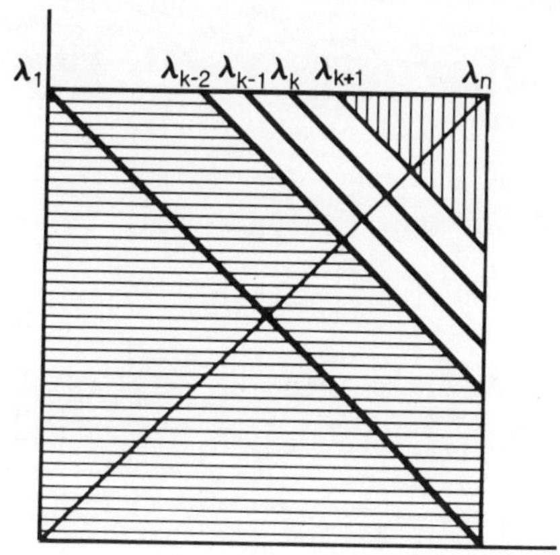

FIGURE 2

If $\underline{p} = (p_1,\ldots,p_n)$ where $n>2$, we denote by $\underline{p}^{ij}$ the vector in $[0,1]^{n-2}$ obtained by deleting the ith and j^{th} coordinates of $\underline{p}$. For any $r \geq 0$, we let $h_r^*(\underline{p}^{ij})$ denote the probability that exactly r of the n-2 components (with respective probabilities given by $\underline{p}^{ij}$) function. Boland and Proschan (1983) prove that

$$h_{k-2}^*(\underline{p}^{ij}) \leq (\geq) \; h_{k-1}^*(\underline{p}^{ij}) \quad \text{for } i \neq j,\ k \geq 2$$

whenever

$$p_\ell \geq \; (\leq) \frac{k-1}{n-1} \quad \text{for all } \ell \neq i,j \; .$$

This result, coupled with the Schur-Ostrowski condition, enable Boland and Proschan to prove the following Schur property of $h_k(\underline{p})$:

Theorem 4. The reliability function $h_k(\underline{p})$: $[0,1]^n \to [0,1]$ is Schur convex in the region $[\frac{k-1}{n-1}, 1]^n$ and Schur concave in the region $[0, \frac{k-1}{n-1}]^n$.

The point $(\frac{k-1}{n-1},\ldots,\frac{k-1}{n-1})$ on the diagonal is a focal point in distinguishing the regions where $h_k(\underline{p})$ is Schur convex and Schur concave. The number $\frac{k-1}{n-1}$ plays an important role in the study of the function of one variable $h_k(p) = \sum_{i=k}^{n} \binom{n}{i} p^i (1-p)^{n-i}$. (Note that $h_k(p)$ is the restriction of $h_k(\underline{p})$ to the diagonal.) $h_k(p)$ is convex on the interval $[0, \frac{k-1}{n-1}]$ and concave on the interval $[\frac{k-1}{n-1}, 1]$ (See Barlow and Proschan (1965)). Figure 3 may help to interpret Theorem 4. Again the vertically (horizontally) shaded region indicates where $h_k(\underline{p})$ is Schur convex (Schur concave). Figure 4 illustrates the combined results of Theorem 3 and Theorem 4.

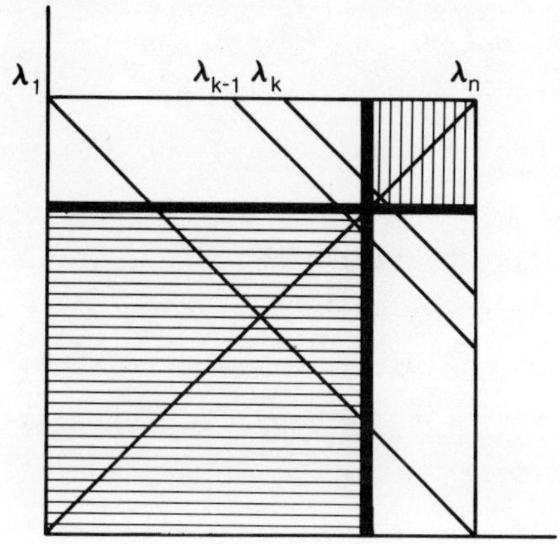

FIGURE 3

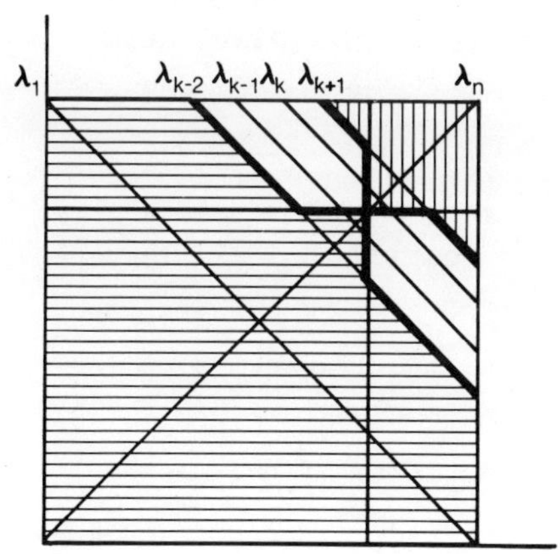

FIGURE 4

III. APPLICATIONS

We now give a number of examples in order to demonstrate how the properties of $h_k(\underline{p})$ presented can help us both in comparing various k out of n systems and in calculating bounds for the reliability of specified k out of n systems.

Example 1. Let us consider a 3 out of 4 system. Theorem 4 in particular implies that a system with component reliabilities (.7, .8, .9, 1.0) is superior (has higher reliability) than a system with component reliabilities (.75, .75, .95, .95) which in turn is superior to one with component reliabilitites (.85, .85, .85, .85). Note that for all three of these systems the four component reliabilities sum to 3.4, and that each component reliability exceeds $\frac{k-1}{n-1} = \frac{2}{3}$. Theorem 4 also implies that the system with component reliabilities (.2, .3, .5, .6) is inferior to one with component reliabilities (.2, .4, .4, .6) which in turn is inferior to one with component reliabilities (.4, .4, .4, .4).

Example 2. We now consider evaluating the reliability of a 5 out of 8 system. As previously indicated, even for this example where k and n are rather small, the standard method for evaluating $h_5(p_1,\ldots,p_8)$ would normally involve (if for example the eight component probabilities are distinct) calculating the sum of 93 products of 8 numbers each. We will now indicate by judiciously 'averaging' some (or all) of the component probabilities, how more easily calculable bounds may be computed for the reliability of the system.

Given a vector $\underline{p} = (p_1,\ldots,p_8)$ of component probabilities, we lose no generality in assuming that $p_1 \le p_2 \le \ldots \le p_8$. Let us use the following notation: $p_{ij}=(p_i+p_j)/2$ and $p_{ij\ell m} = (p_i+p_j+p_\ell+p_m)/4$ for any integers i, j, ℓ, m between 1 and 8.

As before $\overline{p}$ and p_G will denote respectively the arithmetic and geometric means of the components of the vector $\underline{p} = (p_1, \ldots, p_8)$. Note that $\underline{p} = (p_1, p_2, p_3, p_4, p_5, p_6, p_7, p_8) >^m (p_{12}, p_{12}, p_{34}, p_{34}, p_{56}, p_{56}, p_{78}, p_{78})$

$$>^m (p_{1234}, p_{1234}, p_{1234}, p_{1234}, p_{5678}, p_{5678}, p_{5678}, p_{5678})$$

$$>^m (\overline{p}, \overline{p}, \overline{p}, \overline{p}, \overline{p}, \overline{p}, \overline{p}, \overline{p}) = \underline{\overline{p}}$$

The calculation of $h_k(\underline{p})$ becomes easier as the number of distinct component values in $\underline{p}$ decreases. While the calculation of $h_5(\underline{p})$ could involve adding 93 products, the calculations of $h_5(p_{12}, p_{12}, p_{34}, p_{34}, p_{56}, p_{56}, p_{78}, p_{78})$, $h_5(p_{1234}, p_{1234}, p_{1234}, p_{1234}, p_{5678}, p_{5678}, p_{5678}, p_{5678})$ and $h_5(\underline{\overline{p}})$ would involve adding respectively at most 31, 10, and 4 products. If the vector components of $\underline{p} = (p_1, p_2, p_3, p_4, p_5, p_6, p_7, p_8)$ all are $\geq \frac{k-1}{n-1} = \frac{4}{7}$ (respectively $\leq \frac{4}{7}$), then we can easily determine some lower (upper) bounds for $h_5(\underline{p})$ by using Theorems 1 and 4. To illustrate the accuracy of these bounds we compute some specific probabilities.

a) Let $\underline{p} = (.60, .64, .70, .74, .80, .84, .90, .94)$. Each $p_i \geq \frac{4}{7}$ and we observe that

$$\begin{aligned} h_5(\underline{p}) &= .92097 = h_5(.60, .64, .70, .74, .80, .84, .90, .94) \\ &\geq .92069 = h_5(.62, .62, .72, .72, .82, .82, .92, .92) \\ &\geq .91896 = h_5(.67, .67, .67, .67, .87, .87, .87, .87) \\ &\geq .91201 = h_5(.77, .77, .77, .77, .77, .77, .77, .77). \end{aligned}$$

b) Let $\underline{p}$ = (.61, .62, .63, .64, .65, .66, .67, .68). Then again each $p_i \geq \frac{4}{7}$ and we observe that

$$\begin{aligned} h_5(\underline{p}) &= .69580 = h_5(.61, .62, .63, .64, .65, .66, .67, .68) \\ &\geq .69580 = h_5(.615, .615, .635, .635, .655, .655, .675, .675) \\ &\geq .69576 = h_5(.625, .625, .625, .625, .665, .665, .665, .665) \\ &\geq .69562 = h_5(.645, .645, .645, .645, .645, .645, .645, .645) \end{aligned}$$

c) If $\underline{p}$ = (.10, .14, .20, .24, .30, .34, .40, .44), then each component $p_i \leq \frac{4}{7}$ and we observe that

$$\begin{aligned} h_5(\underline{p}) &= .03188 = h_5(.10, .14, .20, .24, .30, .34, .40, .44) \\ &\leq .03206 = h_5(.12, .12, .22, .22, .32, .32, .42, .42) \\ &\leq .03318 = h_5(.17, .17, .17, .17, .37, .37, .37, .37) \\ &\leq .03768 = h_5(.27, .27, .27, .27, .27, .27, .27, .27). \end{aligned}$$

Note that for the given vector $\underline{p}$, the geometric mean of the coordinates is $p_G = .24276$. Using the corollary to Theorem 2 we can compute the lower bound for $h_k(\underline{p})$ which is given by $h_5(\underline{p}_G) = .02410$.

REFERENCES

1. Barlow, R. E., and Proschan, F., "Mathematical Theory of Reliability", John Wiley and Sons, New York, (1965).
2. Boland, P. J., and Proschan, F., The reliability of k out of n systems, *Ann. Prob. 11*, 760-764, (1983).
3. Gleser, L., On the distribution of the number of successes in independent trials, *Ann. Prob. 3*, 182-188, (1975).

4. Hardy, G. H., Littlewood, J. E., and Pólya, G., "Inequalities" (2nd edition), Cambridge University Press, (1952).

5. Hoeffding, W., On the distribution of the number of successes in independent trials, *Ann. Math. Statist. 27*, 713-721, (1956).

6. Marshall, A. W. and Olkin, I., "Inequalities: Theory of Majorization and its Applications", Academic Press, (1979).

7. Pledger, G. and Proschan F., Comparisons of order statistics and of spacings from heterogeneous distributions, In "Optimizing Methods in Statistics" (ed. J.S. Rustagi), 89-113, Academic Press, New York, (1971).

9. Samuels, S. M., On the number of successes in independent trials, *Ann. Math. Statist. 36*, 1272-1278, (1965).

PROXIMITY BETWEEN DISTRIBUTIONS: AN INEQUALITY AND ITS APPLICATIONS

Mark Brown[1]

Department of Mathematics
City College, CUNY
New York, New York

An upper bound is derived for the distance between two probability distributions the ratio of whose survival functions is increasing. Applications to IMRL distributions, NBUE distributions and Laplace transforms are presented. A less general version of the bound was employed by the author in studying the time to first failure for repairable systems.

I. INTRODUCTION

Suppose that F_1 and F_2 are probability distributions with $\overline{F}_1(t)/\overline{F}_2(t)$ increasing in t $(\overline{F}_i(t) = P_{F_i}(X>t),\ i = 1,2)$. Define $d^*(F_1,F_2) = \sup|F_1(B) - F_2(B)|$ as B ranges over the Borel sets on the real line. The following inequality will be derived and applied:

[1] *Research supported by Air Force Office of Scientific Research, under Grant No. AFOSR 82-0024.*

ISBN 0-12-041420-1

$$(1.1) \qquad d^*(F_1,F_2) \le 1 - \int \frac{\overline{F}_2(t^-)}{\overline{F}_1(t^-)}\, dF_1(t)$$

The inequality (1.1) is more general than that developed in Brown ((1979) and (1983a)), and the current proof is simpler. The proof is presented in section 2, and several applications are outlined in section 3.

My interest in this topic was sparked by Keilson's ((1975), (1979)) investigations into the time to first failure for repairable systems. A result of his work was that the key to obtaining bounds for exponential approximations to the distributions of interest was to attack the problem of how much a completely monotone distribution with known first two moments could deviate from an exponential distribution with the same mean. By developing and applying an inequality similar to (1.1), the above problem was solved (Brown (1983a)), and exponential apprxoimations to the time to first failure obtained (Brown (1982), (1983a)).

II. PROOF OF (1.1)

Define $\overline{F}_Z(t) = \overline{F}_2(t)/\overline{F}_1(t)$ and note that $\overline{F}_Z(-\infty) = 1$, $\overline{F}_Z$ is decreasing and right continuous and $0 \le \overline{F}_Z(t) \le 1$ for all t. It is thus the survival function of a possibly improper random variable Z with $\Pr(Z=\infty) = \lim_{t\to\infty} \dfrac{\overline{F}_2(t)}{\overline{F}_1(t)}$. Construct X_1 and Z independent with $X_1 \sim F_1$ and $Z \sim F_Z$. Define $X_2 = \min(X_1,Z)$ and note that $\Pr(X_2>t) = \overline{F}_1(t)\overline{F}_Z(t) = \overline{F}_2(t)$, thus $X_2 \sim F_2$. Now for any Borel set B:

$$|F_1(B) - F_2(B)| = |\Pr(X_1 \in B, X_1 \neq X_2) - \Pr(X_2 \in B, X_1 \neq X_2)|$$

$$\leq \max(\Pr(X_1 \in B, X_1 \neq X_2), \Pr(X_2 \in B, X_1 \neq X_2)) \leq \Pr(X_1 \neq X_2)$$

$$= \Pr(Z < X_1) = 1 - \Pr(Z \geq X_1) = 1 - \int \frac{\overline{F}_2(t^-)}{\overline{F}_1(t^-)} \, dF_1(t)$$

This concludes the proof.

(2.1) *Remark.* If F_1 and F_2 are absolutely continuous then $\overline{F}_1/\overline{F}_2$ increasing is easily shown to be equivalent to the existence of versions of the failure rate functions h_1, h_2 such that $h_1(t) \leq h_2(t)$ for all t. Defining $H_i(t) = -\ln\overline{F}_i(t)$, there are several equivalent expressions for the bound in (1.1):

$$\int \frac{\overline{F}_2(t^-)}{\overline{F}_1(t^-)} \, dF_1(t) = E\left[\frac{\overline{F}_2}{\overline{F}_1}(X_1)\right] = E(H_1(X_2)) = E\left[\frac{h_1}{h_2}(X_2)\right]$$

In the above, $X_1 \sim F_1$ and $X_2 \sim F_2$.

(2.2) *Remark.* If F_1 and F_2 are both concentrated on a countable set S, define the failure rate functions by

$$h_i(x) = \frac{f_i(x)}{\overline{F}_i(x^-)} \quad , \quad i = 1,2, \ x \in S,$$ where f_i is the probability mass function of F_i. The condition that $\overline{F}_1/\overline{F}_2$ be increasing is equivalent to $h_1(x) \leq h_2(x)$ for all $x \in S$. Some equivalent expressions are:

$$\int \frac{\overline{F}_2(t^-)}{\overline{F}_1(t^-)}\, dF_1(t) = \sum_{x \in S} \overline{F}_2(x^-)h_1(x)$$

$$= \sum_{x \in S} \frac{h_1(x)}{h_2(x)}\, f_2(x)$$

III. APPLICATIONS

(3.1) *IMRL Distributions.* Consider an IMRL (increasing mean residual life) distribution F on $[0,\infty)$, defined by $E_F(X-t|X>t)$ increasing in $t \geq 0$. Define $\mu = E_F X$ and $G(x) = \mu^{-1}\int_0^x \overline{F}(s)ds$, the stationary renewal distribution corresponding to F. Note that:

$$\text{(3.1a)} \qquad \frac{\overline{G}(t)}{\overline{F}(t)} = \mu^{-1}E(X-t|X>t)$$

Since F is IMRL, (3.1a) is increasing. Applying (1.1) with $F_1 = G$, $F_2 = F$ we obtain via the Cauchy-Schwartz inequality (Brown (1983a), Theorem 4.1, part (i)):

$$\text{(3.1b)} \qquad d^*(F,G) \leq 1 - \frac{\mu}{\mu_G}$$

In (3.1b) $\mu_G = E_G X = \mu_2/2\mu$, where $\mu_2 = E_F X^2$. Note that for F exponential, $F \equiv G$, and thus $\mu = \mu_G$ and the right hand side of (3.1b) is zero.

Next, let $\mathcal{E}$ be an exponentially distributed random variable with mean 1, and thus $\mu\mathcal{E}$ an exponential with mean μ. Define h_G to be the failure rate function of G and note that:

$$(3.1c) \qquad h_G(t) = \frac{\overline{F}(t)}{\mu\overline{G}(t)} = [E(X-t|X>t)]^{-1}$$

Since F is IMRL, G is DFR (decreasing failure rate). Since $h_G(0) \le \mu^{-1}$ and h_G is decreasing we see that $h_G(t) \le \mu^{-1}$ for all t. Apply (1.1) with $F_1 = G$, and $\mu\mathcal{E} \sim F_2$ to obtain (Brown (1983a), Theorem (4.1) part (iii)):

$$(3.1d) \qquad d^*(G, \mu\mathcal{E}) \le 1 - \frac{\mu}{\mu_G}$$

Thus if F is IMRL and μ is close to μ_G then G is approximately exponential with mean μ.

Note that G is stochastically larger than both F and $\mu\mathcal{E}$. It follows (Brown (1983a), Theorem (4.1) part (v)) that:

$$(3.1e) \qquad \sup|\overline{F}(t) - e^{-t/\mu}| \le \max(d^*(F,G), d^*(G,\mu\mathcal{E})) \le 1 - \frac{\mu}{\mu_G}$$

Thus an IMRL distribution with small $1 - \frac{\mu}{\mu_G} = 1 - \frac{2\mu^2}{\mu_2}$, is approximately exponential. Finally, since $\mu_G\mathcal{E}$ and G are both stochastically greater than $\mu\mathcal{E}$ it follows (Brown (1983a), Theorem (4.1) part (vi)) that:

$$(3.1f) \qquad \sup|\overline{G}(t) - e^{-t/\mu_G}| \le 1 - \frac{\mu}{\mu_G}$$

The inequalities (3.1b) and (3.1e) are sharp even within the subclass of completely monotone distributions.

The motivation for studying approximate exponentiality for completely monotone distributions with small $\rho = \frac{\mu_G}{\mu} - 1$ was provided by Keilson ((1975), (1979)) in his study of first passage time distributions in time reversible Markov processes. Several authors (see Brown (1983a) for references) using Fourier methods derived inequalities for $\sup|\overline{F}(t) - e^{-t/\mu}|$ in terms of ρ. The above results (with bound $\frac{\rho}{\rho+1}$) improve and extend the earlier results. In the reliability context they apply to the problem of time to first failure for coherent systems of independent repairable components (Brown (1982), (1983a)).

(3.2) *NBUE Distributions.* Consider an absolutely continuous distribution L on $[0,\infty)$ with failure rate function h_L. Define m to be the essential infimum of h_L and assume that $m > 0$. Apply (1.1) with $F_1 = c^{-1}\mathcal{E}$ for $0 < c < m$ and $F_2 = L$ to obtain (Brown (1983a), 4.10):

$$\text{(3.2a)} \qquad d^*(L,c^{-1}\mathcal{E}) \le 1 - c\int\overline{L}(t)dt = 1 - c\mu_L$$

Next, consider a NBUE (new better than used) distribution F on $[0,\infty)$. (NBUE is defined by $E(X-t|X>t) \le EX$ for all $t \ge 0$). Let G be the stationary renewal distribution corresponding to F. Then:

$$\text{(3.2b)} \qquad h_G(t) = \frac{\overline{F}(t)}{\mu\overline{G}(t)} = \{E[X-t|X>t]\}^{-1} \ge \mu^{-1}$$

Apply (3.2a) with $L = G$ and $c = \mu^{-1}$ obtaining:

$$(3.2c) \qquad d^*(G, \mu E) \le 1 - \frac{\mu_G}{\mu}$$

Thus within the class of NBUE distributions a small value of $1 - \frac{\mu_G}{\mu}$ implies approximate exponentiality for G. Approximate exponentiality of F itself, for F NBUE and NWUE, is studied by Brown and Ge (1983a).

(3.3) *Laplace Transforms and Measures of Variability.* For a distribution F on $[0,\infty)$ with Laplace transform $L(\alpha) = Ee^{-\alpha X}$, we can consider L to be the survival function of $X^{-1}E$ where $X \sim F$ independent of E. Thus L is the survival function of a completely monotone (and thus IMRL) distribution, and the results of section (3.1) apply. In particular, we use (3.1d) with $\overline{G} = L$, $\mu = -L'(0) = EX$ and $\mu_G = \int L(\alpha)d\alpha = EX^{-1}$, obtaining (Brown (1983b)):

$$(3.3a) \qquad 0 \le L(\alpha) - e^{-\alpha\mu} \le 1 - (EXEX^{-1})^{-1}$$

Using Bernstein's theorem (Feller (1966), p. 415) we can extend (3.3a) to $g \in CM$, where CM is the class of completely monotone functions. Defining $c^2 = 1 - (EXEX^{-1})^{-1}$ we obtain:

$$(3.3b) \qquad 0 \le Eg(X) - g(EX) \le c^2 g(0) \quad \text{for} \quad g \in CM$$

and

$$(3.3c) \qquad \text{Var } g(X) \le c^2 g^2(0) \quad \text{for} \quad g \in CM \ .$$

The interpretation of (3.3c) is that if c^2 is small then the random variable $g(X)/g(0)$ is highly concentrated around $g(EX)/g(0)$. Thus c^2 is a scale invariant measure of variability for X which is reflected in the behavior of completely monotone functions of X.

Note that both $EX^2 \geq (EX)^2$ and $EX^{-1} \geq (EX)^{-1}$ follow from Jensen's inequality, and in both equality holds if and only if X is a constant. The variance σ^2 measures discrepancy between EX^2 and $(EX)^2$ while c^2 measures discrepancy between EX^{-1} and $(EX)^{-1}$.

Define G to be the class of indicator functions of sets of the form $\{|x-\mu| \geq a\}$, so that $g_a(x) = \begin{matrix} 1 \text{ if } |x-\mu| \geq a \\ 0 \text{ otherwise} \end{matrix}$. Defining $A(g_a) = a^{-2}$, Chebichev's inequality can be expressed as:

$$\text{(3.3d)} \qquad 0 \leq Eg(X) - g(EX) \leq \sigma^2 A(g) , \quad g \in G$$

while defining $B(g) = g(0)$, (3.3c) can be expressed as:

$$\text{(3.3e)} \qquad 0 \leq Eg(X) - g(EX) \leq c^2 B(g) , \quad g \in CM .$$

In both cases, bounds for bias are obtained for a class of functions. In (3.3d) the measure of variability, σ^2, appears while c^2 appears in (3.3e). Both have functionals appearing on the right side, $A(g) = a^{-2}$ for $g \in G$ in (3.3d) and $B(g)=g(0)$ for $g \in CM$ in (3.3e).

(3.4) *Occurrence of Patterns.* In Brown and Ge (1983b) (appearing in this volume) inequality (1.1) is used to approximate the distribution of the waiting time for the first occurrence of a specified pattern in a sequence of multinomial trials.

IV. FURTHER REMARK

It can also be proved that if F_1 and F_2 are probability distributions with $F_1(t)/F_2(t)$ increasing then:

$$(4.1) \qquad d^*(F_1,F_2) \le 1 - \int \frac{F_1(t)}{F_2(t)}\, dF_2(t)$$

Although (4.1) appears to be potentially useful, I have not yet found any applications of interest.

REFERENCES

1. Brown, M., "Approximating DFR distributions by exponential distributions, with applications to first passage times", AFOSR Technical Report M517, FSU Technical Report No. 79-B2, August, (1979).
2. Brown, M., "On the reliability of repairable systems", AFOSR Technical Report No. 82-01, City College, CUNY, Report No. MB1, October, 1982, to appear in *Operations Research*.
3. Brown, M., "Approximating IMRL distributions by exponential distributions, with applications to first passage times", *Ann. Probability*, Vol. 11, No. 2, 419-427, (1983a).
4. Brown, M., "A measure of variability based on the harmonic mean and its use in approximation", AFOSR Technical Report No. 82-03, City College, CUNY Report No. MB3, March, (1983b).
5. Brown, M., and Ge, G., "Exponential approximations for two classes of aging distributions", AFOSR Technical Report No. 82-02, City College, CUNY Report No. MB2, February, 1983a, to appear in the *Annals of Probability*.
6. Brown, M, and Ge, G., "On the waiting time for the first occurrence of a pattern", to appear in this volume, (1983b).

7. Feller, W., *An Introduction to Probability Theory and its Applications, Volume II,* John Wiley, New York, (1966).

8. Keilson, J., "Systems of independent Markov components and their transient behavior", *Reliability and Fault Tree Analysis,* (R.E. Barlow, ed.), SIAM, Philadelphia, 351-364, (1975).

9. Keilson, J., *Markov Chain Models: Rarity and Exponentiality,* Springer-Verlag, New York, (1979).

ON THE WAITING TIME FOR THE FIRST OCCURRENCE OF A PATTERN

Mark Brown[1]

Department of Mathematics
City College, CUNY
New York, New York

Guangping Ge

Department of Mathematics
Hebei Teachers University
Shijiazhuang, Hebei
The Peoples Republic of China

We look at the waiting time for the first occurrence of a pattern in multinomial trials. Using reliability theory methodology it is shown that a truncated geometric distribution provides a simple and highly accurate approximation for nonrepeating rare patterns.

[1] *Research supported by the Air Force Office of Scientific Research under AFOSR Grant No. 82-0024.*

ISBN 0-12-041420-1

I. INTRODUCTION

It has often been said that a monkey randomly banging away at a typewriter will eventually reproduce the complete works of Shakespeare. This is one example of the problem of interest in this paper. Given a sequence of i.i.d. multinomial trials, find the distribution of the waiting time for the first occurrence of a specified pattern.

Using methodology of reliability theory we obtain a simple but highly accurate approximation. For a non-repetitive pattern (defined in section 3) of length m and probability p we employ the following intuitively based approximation to the distribution of T, the waiting time to the first occurrence. First note that $T > m - 1$ and $\Pr(T=m) = p$. We would anticipate that for $K \geq m$, $\Pr(T=K|T \geq K) \approx p$. This suggests using an approximating distribution with failure rate 0 for $K<m$ and failure rate p for $K \geq m$, i.e. the conditional distribution of X given $X \geq m$ where X is geometrically distributed with parameter p. For this approximation we obtain in Theorem (3.1):

$$(1.1) \qquad 0 \leq q^{K-(m-1)} - \Pr(T>K) \leq (m-1)p \quad \text{for} \quad K \geq m-1$$

$$(1.2) \qquad \sup_B \left| \sum_B q^{K-m} p - \Pr(T \in B) \right| \leq (m-1)p$$

where in (1.2) B ranges over all subsets of $\{m, m+1, \ldots\}$.

Feller (1967) chapter 16, derives the exact distribution of T, but its computation requires the complete spectral analysis of an $n^m \times n^m$ transition matrix (where n is the number of possible outcomes for X). As an approximation Feller suggests $\Pr(T>K) \approx Ab^K$ where b is the largest eigenvalue less than 1, and also coincides with 1 minus the limiting failure rate of T. This approximation is good for large K but it is not clear how accurate it is on a global basis. The exact distribution can also be obtained by repeated differentiation of the probability generating function and by stagewise solution of recursive equations. Both methods are cumbersome, and require considerable computational effort for large K.

II. DISCRETE NBUE DISTRIBUTIONS

Consider a discrete random variable, X, taking values in $\{1,2,\ldots\}$. Define $F(K) = \Pr(X=K)$, and $\overline{F}(K) = \Pr(X>K)$. The stationary renewal distribution, G, corresponding to F has probability mass function $g(x) = \overline{F}(x)/\mu$ where $\mu = EX$ (see Feller (1967), chapter 13, for a discussion of discrete renewal processes). Let X^* denote a random variable with distribution G. Define F to be NBUE (new better than used in expectation) if $E(X-t|X>t)\leq\mu$ for $t = 1,2,\ldots$. Since:

$$(2.1)\quad E(X-t|X>t) = \frac{\mu\overline{G}(t-1)}{\overline{F}(t)}$$

it follows that F NBUE is equivalent to X being stochastically greater than $X^* + 1$.

For a discrete random variable, X, the failure rate function is defined by $h(x) = \Pr(X=x|X\geq x)$. The failure rate function of G is thus:

$$(2.2)\qquad h_G(t) = \frac{\overline{F}(t)}{\mu\overline{G}(t-1)} = [E(X-t|X>t)]^{-1}$$

It follows from (2.2) that F NBUE is equivalent to $h_G(t)\geq\mu^{-1}$ for all t .

The following lemma is due to Brown (1983) and generalizes a similar result (Brown (1979)) which is useful in bounding the distance between distributions.

Lemma 2.1 Let F_1, F_2 be two probability distributions with $\overline{F}_1(t)/\overline{F}_2(t)$ increasing in t. Define $d^*(F_1,F_2) = \sup|F_1(B) - F_2(B)|$ as B ranges over the Borel sets on the real line. Then:

$$d^*(F_1,F_2) = 1 - \int \frac{\overline{F}_2(t^-)}{\overline{F}_1(t^-)} \, dF_1(t)$$

If F_1 and F_2 are discrete distributions taking values in $\{0,1,\ldots\}$ then $\overline{F}_1/\overline{F}_2$ increasing is equivalent to $h_1(x) \le h_2(x)$ for $x = 0,1,2,\ldots$. Returning to F NBUE we saw (2.2) that $h_G(x) \ge \mu^{-1}$. The constant $h \equiv \mu^{-1}$ is the failure rate function corresponding to the geometric distribution $G_{\mu^{-1}}$, where:

$$(2.3) \qquad \Pr(G_{\mu^{-1}} = K) = \mu^{-1}(1-\mu^{-1})^K, \; K = 0,1,\ldots, \quad .$$

We can therefore apply *Lemma 2.1* with $F_1 = G_{\mu^{-1}}$ and $F_2 = G$ to obtain:

$$(2.4) \qquad d^*(G_{\mu^{-1}},G) \le 1 - \mu^{-1} \sum_0^\infty \overline{G}(K^-) = 1 - \frac{\mu_G+1}{\mu} \quad .$$

Since $G_{\mu^{-1}}$ is stochastically larger than G, it follows from (2.4) that:

$$(2.5) \qquad 0 \le (1-\mu^{-1})^{K+1} - \overline{G}(K) \le 1 - \frac{\mu_G+1}{\mu} \quad \text{for} \quad K = 0,1,2,\ldots \quad .$$

III. PATTERNS

Consider a sequence of i.i.d. random variables $\{X_1,X_2,\ldots\}$ taking values in a countable set S, with $\Pr(X=s) = p(s) > 0$ for $s \in S$. Let $a_1,\ldots,a_m$ be an ordered m-tuple with each $a_i \in S$.

Define $p = Pr(a_1,\ldots,a_m) = \prod_{i=1}^{m} p(a_i)$. The ordered array $a_1,\ldots,a_m$ will be called a pattern. A non-repetitive pattern is one which for each $K = 1,\ldots,m-1$, $a_{m-K+1} \ldots a_m$ does not coincide with $a_1,\ldots,a_K$. For example ABC and 1113 are non-repetitive while ABCA and 22 are repetitive.

Define T to be the waiting time for the first occurrence of the non-repetitive pattern $a_1,\ldots,a_m$. Our main theorem now follows:

Theorem 3.1. For a non-repetitive pattern of length m and probability p:

(i) $0 \leq q^{K-(m-1)} - Pr(T>K) \leq (m-1)p$ for $K \geq m-1$, where $q = 1-p$.

(ii) $\sup \left| \sum_{K\in B} q^{K-m}p - Pr(T\in B) \right| \leq (m-1)p$ where B ranges over the subsets of $\{m,m+1,\ldots\}$.

Proof. Define T_K to be the waiting time starting at time K for the first occurrence of the pattern at or after time K. The distribution of T_K is the same for all $K \geq m$, and it thus coincides with T^*, where T^* is distributed as the stationary renewal distribution corresponding to T. Since $T = m+T_m$ and $T_m \sim T^*$ it follows that:

$$(3.1) \qquad T^* \sim T - m; \quad \mu_G = ET^* = \frac{1}{p} - m$$

Since $T^* \sim T - m \leq T - 1$, T is NBUE. By elementary renewal theory, $ET = \frac{1}{p}$. It thus follows from (2.4) and (3.1) that:

$$d^*(G_p, T-m) \leq 1 - \frac{\mu_G + 1}{\mu} = 1 - p\left(\frac{1}{p} - (m-1)\right) = (m-1)p.$$

This proves (i); (ii) follows from (2.5).

REFERENCES

1. Brown, M. (1979). "Approximating DFR distributions by exponential distributions, with applications to first passage times." *AFOSR Technical Report M517, FSU Technical Report No. 79-B2*, August 1979.
2. Brown, M. (1983). "Proximity between distributions: An inequality and its applications." To appear in this volume.
3. Feller, W. (1967). *An Introduction to Probability Theory and its Applications, Volume I, 3rd Edition.* John Wiley, New York.

A SUMMARY OF RESULTS CONCERNING DEPENDENCE PROPERTIES PRESERVED UNDER MULTIPLE INTEGRATIONS

Mei-Ling Ting Lee

Bridgewater State College
Bridgewater, Massachusetts

Notions of positive dependence have arisen from the consideration of a TP_2 property of a multiple integration kernels. Results of this type have been considered in the bivariate case by Shaked (1977) and in the multivariate case by Lee (1983a). Using a negative dependence concept of Karlin and Rinott (1980) as a counterpart for TP_2 in pairs, a negative dependence analogue was considered by Lee (1983b). In this paper these results are reviewed and summarized.

0. INTRODUCTION

Shaked (1977) considered a concept of dependence motivated by reliability theory. He called this dependence by total positivity (DTP). In section 1, we briefly review Shaked's results. A multivariate generalization is given in section 2. Although the extension of the definition seems natural, results and proofs are not immediate. In section 3, we give a multivariate analogue of the families considered by Yanagimoto (1972), and discuss its relationships with the families

ISBN 0-12-041420-1

considered in section 2. Other dependence concepts considered by Barlow and Proschan (1975) are also discussed in section 3. A negative dependence analogue is introduced in section 4. Properties and characterizations through reliability functions are summarized.

1. THE BIVARIATE CASE

Dependence properties preserved under multiple integration were first discussed by Shaked (1977). Let (X,Y) be a random vector with joint density function f(x,y). Observing that the conditional mean residual life function

$$m(x|Y = y) = E(X-x|X > x,\ Y = y)$$

is increasing in y for every x if and only if

$$\int_x^\infty \int_{x_1}^\infty f(x_0,y)\,dx_0 dx_1$$

is TP_2 in x and y, and that the conditional hazard rate function

$$\gamma(x|Y > y) = \frac{f(x|Y > y)}{\bar{F}(x|Y > y)}$$

is decreasing in y for every x if and only if

$$\int_y^\infty \int_x^\infty f(x_0,y_0)\,dx_0 dy_0$$

is TP_2 in x and y, Shaked considered the multiple integration

$$\Psi_{m,n}(x,y) = \int_y^\infty \int_{y_{n-1}}^\infty \cdots \int_{y_1}^\infty \int_x^\infty \int_{x_{m-1}}^\infty \cdots \int_{x_1}^\infty f(x_0,y_0)$$

$$dx_0 \cdots dx_{m-1} dy_0 \cdots dy_{n-1}$$

and defined the concept that the random vector (X,Y) is dependent by total positivity of order two with degree (m,n), denoted by DTP(m,n), if $\Psi_{m,n}(x,y)$ is TP_2 in x and y $(-\infty < x, y < \infty)$.

Let (X,Y) be an absolutely continuous random vector with the support of X denoted by S_X and the support of Y denoted by S_Y. Shaked (1977) showed that (X,Y) is DTP(n,0) if and only if

$$\frac{E[(X-x)^{n-1} | X > x, Y = y]}{E[(X-x)^{n-2} | X > x, Y = y]}$$

is increasing in $y \in S_Y$ for every $x \in S_X$, and that (X,Y) is DTP(n,1) if and only if

$$\frac{E[(X-x)^{n-1} | X > x, Y > y]}{E[(X-x)^{n-2} | X > x, Y > y]}$$

is increasing in $y \in S_Y$ for all $x \in S_X$, provided the expectations exists and $n > 1$.

Several interrelationships for DTP random variables were derived as were some moment inequalities. Assume (X,Y) is DTP(m,n), $m \geq 0$, $n \geq 0$, then (X,Y) is DTP(s_1,s_2) for every $s_1 \in \{m\} \cup [m+1, \infty)$ and $s_2 \in \{n\} \cup [n+1, \infty)$. Also under the same assumption, $\Psi_{s_1,s_2}(x,y)$ is TP_2 in s_1 and s_2 for $s_1 \geq m$, $s_2 \geq n$.

Shaked proved that for m,n = 0,1,2, if (X,Y) is DTP(m,n), then $\text{cov}(X,Y) \geq 0$, provided it is defined. And for m,n = 0,1, it was shown that the families of DTP(m,n) distributions coincide with some families of positive dependence distributions introduced by Yanagimoto (1972). Relationships with other concepts of dependence discussed in Barlow and Proschan (1975) (see Figure 5.4.1, p. 146) were shown in the following chart of implications.

$$\mathrm{DTP}(0,0) \longrightarrow \mathrm{DTP}(1,0) \begin{array}{l} \nearrow \mathrm{DTP}(2,0) \searrow \\ \rightarrow \mathrm{RCSI} \longrightarrow \mathrm{DTP}(2,1) \\ \searrow \mathrm{SI} \quad \searrow \mathrm{RTI} \longrightarrow \text{Association.} \end{array}$$

These results will be discussed in more detail in section 3.

2. DEFINITION AND PROPERTIES OF MULTIVARIATE DTP FAMILIES

To extend the concepts of Shaked to the multivariate case we proceed as follows. Let

$$\gamma^{(s)}(t) = \begin{cases} (-t)^{s-1}/\Gamma(s), & t \le 0 \\ 0 & , t > 0 \end{cases}$$

and let $X_1,\ldots,X_n$ be random variables with joint distribution function F.

For $k_i > 0$, define the n fold integral $\Psi_{k_1,\ldots,k_n}(x_1,\ldots,x_n)$ by

$$\Psi_{k_1,\ldots,k_n}(x_1,\ldots,x_n) = \int_{-\infty}^{\infty}\cdots\int_{-\infty}^{\infty}\gamma^{(k_1)}(x_1-t_1)\cdots\gamma^{(k_n)}(x_n-t_n)dF$$

$(t_1,\ldots,t_n)$ and define $\Psi_{0,\ldots,0}(x_1,\ldots,x_n) = f(x_1,\ldots,x_n)$ if the joint density exists.

Also define $\Psi_{0,\ldots,0,k_{i+1},\ldots,k_n}(x_1,\ldots,x_n)$ to be the (n-i) fold integral

$$\int_{-\infty}^{\infty}\cdots\int_{-\infty}^{\infty}\gamma^{(k_{i+1})}(x_{k_{i+1}}-t_{i+1})\cdots\gamma^{(k_n)}(x_{k_n}-t_n)g(x_1,\ldots,x_i)$$
$$dF(t_{i+1},\ldots,t_n|x_1,\ldots,x_i)$$

where g is the joint density of $X_1,\ldots,X_i$, and $F(t_{i+1},\ldots,t_n|x_1,\ldots,x_i)$ is the conditional distribution of

$x_{i+1},\ldots,X_n$ given $X_1 = x_1,\ldots,X_i = x_i$, for $k_{i+1} > 0,\ldots,k_n > 0$. Similarly, we can define $\Psi_{k_1,\ldots,k_n}(x_1,\ldots,x_n)$ with any subset of $\{k_1,\ldots,k_n\}$ being zero.

We define multivariate distributions to be dependent by total positivity as follows. $(X_1,\ldots,X_n)$ is said to be dependent by total positivity of order 2 with degree $(k_1,\ldots,k_n)$, denoted by $DTP(k_1,\ldots,k_n)$, if $\Psi_{k_1,\ldots,k_n}(x_1,\ldots,x_n)$ is TP_2 in pairs of $x_1,\ldots,x_n$.

For example, if (X_1,X_2,X_3) is DTP(1,1,1), then $\int_{x_1}^{\infty}\int_{x_2}^{\infty}\int_{x_3}^{\infty} dF(t_1,t_2,t_3)$ is TP_2 in pairs of x_1,x_2 and x_3.

DTP random variables satisfy some desirable multivariate properties. The joint distribution of two independent sets of DTP random variables is DTP. DTP families are closed under linear transformations; that is, if $(X_1,\ldots,X_n)$ is $DTP(k_1,\ldots,k_n)$, then $(a_1X_1+b_1,\ldots,a_nX_n+b_n)$ is $DTP(k_1,\ldots,k_n)$, for any $a_1,\ldots,a_n > 0$ and $b_1,\ldots,b_n$ real. Under certain conditions, a subset of DTP random variables is DTP. Assume $(X_1,\ldots,X_n)$ is $DTP(k_1,\ldots,k_n)$, with $k_i = 0$ or 1 for some $1 \leq i \leq n$, then $(X_1,\ldots,X_{i-1},\ldots,X_n)$ is $DTP(k_1,\ldots,k_{i-1},k_{i+1},\ldots,k_n)$. See Lee (1983a) for details.

One of the major results in Lee (1983a) is that DTP of a fixed order implies DTP of higher orders. As a corollary, if a joint density function is TP_2 in pairs, then it implies that the joint survival function is TP_2 in pairs.

For some DTP families, pairs of random variables have nonnegative covariance. If $(X_1,\ldots,X_n)$ is $DTP(k_1,\ldots,k_n)$ with $k_i = 0$ or 1 for $i = 1,\ldots,n$, then $\text{cov}(X_i,X_j) \geq 0$ for any

$i \neq j$, $i,j = 1,\ldots,n$, provided it is defined. One counterexample can be found in Lee (1983a) that there exists random variables X,Y and Z such that (X,Y) is DTP(0,0), (Y,Z) is DTP(0,0), and (X,Z) is DTP(0,0), but (X,Y,Z) is not DTP(0,0,0).

Some examples of the DTP families can be constructed easily. Let $T_1,T_2,\ldots,T_n$ be i.i.d. nonnegative random variables with a common density function f. If f is a Polya frequency function of order 2, then the joint distribution of $X_j = \sum_{i=1}^{j} T_i, j = 1,\ldots,n$ is DTP(0,...,0). Let $(Y_1,\ldots,Y_k)$ be RCSI, and let $S_i = \min_{j \in J_i} Y_j$, $i=1,\ldots,n$, where J_i is any subset of $\{1,\ldots,k\}$. Then $(S_1,\ldots,S_n)$ is DTP(1,...,1). Also, if random variables $U_1,\ldots,U_n$ are conditionally independent given W = w, and if (U_i,W) is DTP$(k_i,0)$ for $i = 1,\ldots,n$, then $(U_1,\ldots,U_n)$ is DTP$(k_1,\ldots,k_n)$. By using this multivariate analogue, several examples given in Shaked can also be generalized.

3. RELATIONSHIPS WITH RELIABILITY CONCEPTS.

First, we summarize the relationships with the multivariate positive dependence concepts discussed in Barlow and Proschan (1975), and Block and Ting (1981). In the bivariate case, a random vector is DTP(1,1) if and only if it is RCSI. The multivariate generalization of the "if" part holds. The "only if" part, however, does not hold without a condition on the support of the joint survival function. We have the result that if $(X_1,\ldots,X_n)$ is DTP(1,...,1) and such that $\overline{F}(x_1,\ldots,x_n)$ takes positive values on a product space, then $(X_1,\ldots,X_n)$ is RCSI. For nonnegative random vector $\underset{\sim}{X} = (X_1,\ldots,X_n)$, $\underset{\sim}{X}$ is DTP(0,...,0,1) implies $\underset{\sim}{X}$ is CIS; and $\underset{\sim}{X}$ is DTP(1,...,1) implies $\underset{\sim}{X}$ is RTIS.

Yanagimoto (1972) defined families of positively dependent bivariate distributions $P(i,j)$, by considering four two dimensional intervals. The random vector (X,Y)

(1) is said to be $P(3,3)$ if

$$\frac{P(a_1<X\leq a_2,\ b_2<Y\leq b_3)}{P(a_1<X\leq a_2,\ b_1<Y\leq b_2)} \leq \frac{P(a_2<X\leq a_3,\ b_2<Y\leq b_3)}{P(a_2<X\leq a_3,\ b_1<Y\leq b_2)}$$

for all $a_1<a_2<a_3$, $b_1<b_2<b_3$ such that the denominators do not vanish;

(2) is said to be in $P(2',3)$ if

$$\frac{P(a_1<X\leq a_2,\ b_2<Y\leq b_3)}{P(a_1<X\leq a_2,\ b_1<Y\leq b_2)} \leq \frac{P(a_2<X,\ b_2<Y\leq b_3)}{P(a_2<X,\ b_1<Y\leq b_2)}$$

for all $a_1<a_2$, $b_1<b_2<b_3$ such that the denominators do not vanish;

(3) is said to be in $P(2',2')$ if

$$\frac{P(a_1<X\leq a_2,\ b_2<Y)}{P(a_1<X\leq a_2,\ b_1<Y\leq b_2)} \leq \frac{P(a_2<X,\ b_2<Y)}{P(a_2<X,\ b_1<Y\leq b_2)},$$

for all $a_1<a_2$, $b_1<b_2$ such that the denominators do not vanish.

Shaked showed that if (X,Y) is absolutely continuous, then (X,Y) is DTP(0,0) iff (X,Y) is $P(3,3)$; (X,Y) is DTP(0,1) iff (X,Y) is $P(3,2')$; and (X,Y) is DTP(1,1) iff (X,Y) is $P(2',2')$.

We consider a multivariate analogue of Yanagimoto's families of dependent d.f.'s. Notice that Block, Savits and Shaked (1982) defined a probability measure μ on $\mathbb{R}^n$ to be TP_2 in pairs if $\mu(I_1,I_2,\ldots,I_n)$ is TP_2 in pairs of I_i,I_j for all $i,j = 1,\ldots,n$, and for all one dimensional intervals $I_1 < I_1',\ldots,I_n < I_n'$, where the notation $I < I'$ means $x < y$ for any $x \in I$, $y \in I'$. With this idea, but using specific types of the intervals, we have the following definitions.

Let $\underset{\sim}{X} = (X_1,\ldots,X_n)$ be a random vector with probability measure μ, then

(1) $\underset{\sim}{X}$ is said to be $P(3,3,\ldots,3)$ if $\mu(I_1,I_2,\ldots,I_n)$ is TP_2 in pairs of I_i,I_j for all $1 \leq i \neq j \leq n$, with $I_k = (a_{k_1},a_{k_2})$, $I'_k = (a_{k_2},a_{k_3})$ for any $a_{k_1} < a_{k_2} < a_{k_3}$, $k = 1,2,\ldots,n$.

(2) $\underset{\sim}{X}$ is said to be $P(3,\ldots,3,2',3,\ldots,3)$, with $2'$ in the ℓ-th coordinate, if $\mu(I_1,I_2,\ldots,I_n)$ is TP_2 in pairs of I_i,I_j for all $1 \leq i \neq j \leq n$, with $I_\ell = (a_{\ell_1},a_{\ell_2})$, $I'_\ell = (a_{\ell_2},\infty)$ for any $a_{\ell_1} < a_{\ell_2}$, and $I_k = (a_{k_1},a_{k_2})$, $I'_k = (a_{k_2},a_{k_3})$ for any $a_{k_1} < a_{k_2} < a_{k_3}$, and $k \neq \ell$.

(3) $\underset{\sim}{X}$ is said to be $P(2',2',\ldots,2')$ if $\mu(I_1,I_2,\ldots,I_n)$ is TP_2 in pairs of I_i,I_j for all $1 \leq i \neq j \leq n$, with $I_k = (a_{k_1},a_{k_2})$, $I'_k = (a_{k_2},\infty)$ for any $a_{k_1} < a_{k_2}$ and $k = 1,2,\ldots,n$.

If μ is TP_2 in pairs according to Block et al (1982), it implies that $\underset{\sim}{X}$ is $P(3,\ldots,3)$. Conversely, if $\underset{\sim}{X}$ is $P(3,\ldots,3)$, then μ is TP_2 in the sense of Block et al. Relationships of the multivariate DPT families to the generalized Yanagimoto's families of dependent distributions are stated as follows.

(a) Let $\underset{\sim}{X}$ be a random vector with a density f with respect to a product measure $m = m_1\times\ldots\times m_n$ of σ-finite measures such that f is continuous on the support of m (denoted by S) and zero off S, then $\underset{\sim}{X}$ is $P(3,\ldots,3)$ implies that $\underset{\sim}{X}$ is $DTP(0,\ldots,0)$. If we assume that $\{f > 0\} \cap S = \hat{S}$ is a product space, and f is TP_2 in pairs on $\hat{S}$, then the converse holds, i.e. $\underset{\sim}{X}$ is $DTP(0,\ldots,0)$ implies that $\underset{\sim}{X}$ is $P(3,\ldots,3)$.

(b) Let $(X_1,\ldots,X_n)$ be absolutely continuous. If $\underset{\sim}{X}$ is $P(3,\ldots,3,2',3,\ldots,3)$ then $\underset{\sim}{X}$ is DTP$(0,\ldots,0,1,0,\ldots,0)$.

(c) If $\underset{\sim}{X}$ is $P(2',2',\ldots,2')$ then X is DTP$(1,1,\ldots,1)$.

Proof

(a) See Block et al (1982).

(b) Without loss of generality, let $J_3,J_4,\ldots,J_n$ be arbitrary bounded intervals, then

$$\mu(I_1,I_2,J_3,\ldots,J_n)\mu(I_1',I_2',J_3,\ldots,J_n) \geq \mu(I_1',I_2,J_3,\ldots,J_n)\mu(I_1,I_2',J_3,\ldots,J_n)$$

where $I_1 = (a_{11},a_{12})$, $I_1' = (a_{12},\infty)$, $I_2 = (a_{21},a_{22})$, $I_2' = (a_{22},a_{23})$ for any $a_{11} < a_{12}$, $a_{21} < a_{22} < a_{23}$ real. Adding $\mu(I_1',J_2,J_3,\ldots,J_n)\mu(I_1',I_2',J_3,\ldots,J_n)$ to both sides of the above inequality, we get

$$\mu((a_{11},\infty),I_2,J_3,\ldots,J_n)\mu((a_{12},\infty),I_2',J_3,\ldots,J_n) \geq \mu((a_{12},\infty),I_2,J_3,\ldots,J_n)\mu((a_{11},\infty),I_2',J_3,\ldots,J_n)$$

for any $a_{11} < a_{12}$, $a_{21} < a_{22} < a_{23}$, and $J_3,\ldots,J_n$ bounded intervals.

Hence

$$\int_{x_1}^{\infty} f(t,x_2,x_3,\ldots,x_n)dt \int_{x_1'}^{\infty} f(t,x_2',x_3,\ldots,x_n)dt$$

$$\geq \int_{x_1'}^{\infty} f(t,x_2,x_3,\ldots,x_n)dt \int_{x_1}^{\infty} f(t,x_2',x_3,\ldots,x_n)dt.$$

Similarly, $\int_{x_1}^{\infty} f(t,x_2,x_3,\ldots,x_n)dt$ is TP_2 in x_1 and x_j, $j = 3,\ldots,n$. Now, for any I_1' fixed, $\mu(I_1',J_2,J_3,\ldots,J_n)$ is TP_2 in pairs of $J_2,\ldots,J_n$, thus it follows that

$\int_x^\infty f(t,x_2,\ldots,x_n)dt$ is TP_2 in pairs of $x_2,\ldots,x_n$ for any x_1 fixed. Hence $\Psi_{1,0,\ldots,0}(x_1,\ldots,x_n)$ is TP_2 in pairs of $x_1,\ldots,x_n$ or $\underset{\sim}{X}$ is DTP(1,0,...,0).

(c) Without loss of generality, we show that $P(X_1 > x_1, X_2 > x_2, \ldots, X_n > x_n)$ is TP_2 in x_1,x_2 for any $x_3,\ldots,x_n$ fixed. By the assumption that $\underset{\sim}{X}$ is $\mathcal{P}(2',\ldots,2')$, we have

$$\mu((x_1,x_1'),(x_2,x_2'),(x_3,\infty),\ldots,(x_n,\infty))\,\mu((x_1',\infty),(x_2',\infty),(x_3,\infty),\ldots,(x_n,\infty))$$
$$\geq \mu((x_1,x_1'),(x_2',\infty),(x_3,\infty),\ldots,(x_n,\infty))\,\mu((x_1',\infty),(x_2,x_2'),(x_3,\infty),\ldots,(x_n,\infty))$$

Adding

$$\mu((x_1',\infty),(x_2',\infty),(x_3,\infty),\ldots,(x_n,\infty))\,\mu((x_1,x_1'),(x_2',\infty),(x_3,\infty),\ldots,(x_n,\infty))$$

to both sides of the inequality we get

$$\mu((x_1,x_1'),(x_2,\infty),(x_3,\infty),\ldots,(x_n,\infty))\,\mu((x_1',\infty),(x_2',\infty),(x_3,\infty),\ldots,(x_n,\infty))$$
$$\geq \mu((x_1,x_1'),(x_2',\infty),(x_3,\infty),\ldots,(x_n,\infty))\,\mu((x_1',\infty),(x_2,\infty),(x_3,\infty),\ldots,(x_n,\infty))$$

Again adding

$$\mu((x_1',\infty),(x_2',\infty),(x_3,\infty),\ldots,(x_n,\infty))\,\mu((x_1',\infty),(x_2,\infty),(x_3,\infty),\ldots,(x_n,\infty))$$

to both sides of the inequality, the result follows.

We now summarize the relationships with reliability theory. For absolutely continuous random vector $\underset{\sim}{X}$ with d.f. F and density f, let $S_{\underset{\sim}{X}} = \{\underset{\sim}{x}: f(\underset{\sim}{x}) > 0\}$. Denote $\underset{\sim}{X}^{(i)} = (X_1,\ldots,X_{i-1}, X_{i+1},\ldots,X_n)$. Let A be a Borel set in $S_{\underset{\sim}{X}^{(i)}}$, consider the conditional hazard rate defined by

$$r(x_i|\underset{\sim}{X}^{(i)} \varepsilon A) = \frac{f(x_i|\underset{\sim}{X}^{(i)} \varepsilon A)}{\overline{F}(x_i|\underset{\sim}{X}^{(i)} \varepsilon A)},$$

and the mean residual life function defined by $m(x_i|\underset{\sim}{X}^{(i)} \varepsilon A) =$

$$\int_{x_i}^{\infty} \frac{\overline{F}(t|\underset{\sim}{X}^{(i)} \varepsilon A)}{\overline{F}(x_i|\underset{\sim}{X}^{(i)} \varepsilon A)} dt = E(X_i - x_i|X_i > x_i, \underset{\sim}{X}^{(i)} \varepsilon A).$$

Then the following monotonicity properties hold.

(1) If $\underset{\sim}{X}$ is DTP(0,0,...,0,1) then $r(x_n|\underset{\sim}{X}^{(n)} = \underset{\sim}{x}^{(n)})$ is decreasing in $\underset{\sim}{x}^{(n)} \varepsilon S_{\underset{\sim}{X}^{(n)}}$, for any x_n.

(2) If $\underset{\sim}{X}$ is DTP(0,0,...,0,2), then $m(x_n|\underset{\sim}{X}^{(n)} = \underset{\sim}{x}^{(n)})$ is decreasing in $\underset{\sim}{x}^{(n)} \varepsilon S_{\underset{\sim}{X}^{(n)}}$, for any x_n.

(3) If $\underset{\sim}{X}$ is DTP(0,0,...,0,m), for m > 1, then

$$\frac{E[(X_n - x_n)^{m-1}|X_n > x_n, \underset{\sim}{X}^{(n)} = \underset{\sim}{x}^{(n)}]}{E[(X_n - x_n)^{m-2}|X_n > x_n, \underset{\sim}{X}^{(n)} = \underset{\sim}{x}^{(n)}]}$$

is increasing in $\underset{\sim}{x}^{(n)} \varepsilon S_{\underset{\sim}{X}^{(n)}}$, for any x_n.

(4) $\underset{\sim}{X}$ is DTP(1,...,1) if and only if $r(x_j|\underset{\sim}{X}^{(j)} > \underset{\sim}{x}^{(j)})$ is decreasing in $\underset{\sim}{x}^{(j)}$, for any x_j, j = 1,2,...,n.

By using property (4) of the above results, it can be shown that in the multivariate case, if $\underset{\sim}{X}$ is IHR (Harris) then it is IHR (Johnson-Kotz).

4. NEGATIVE DEPENDENCE ANALOGUE

Since negative dependence concepts are not simply mirror image of positive dependence properties (except in the bivariate case), results are not immediate (see Block, Savits and Shaked (1982)). We use a specialized version of a reverse TP_2 concept due to Karlin and Rinott (1980).

Karlin and Rinott (1980) define a nonnegative function satisfying the property

$$f(\underset{\sim}{x} \wedge \underset{\sim}{y}) f(\underset{\sim}{x} \vee \underset{\sim}{y}) \leq f(\underset{\sim}{x}) f(\underset{\sim}{y}) \quad \text{for all } \underset{\sim}{x}, \underset{\sim}{y}$$

to be multivariate reverse regular of order 2 (MRR_2). An MRR_2 function $f(\underset{\sim}{x})$ is said to be strongly-MRR_2 (S-MRR_2) if for any set of PF_2 functions $\{\phi_\nu\}$, each resulting marginal

$$g(x_{\nu_1}, \ldots, x_{\nu_k}) = \int \ldots \int f(x_1, \ldots, x_n) \phi_1(x_{j_1}) \ldots \phi_{n-k}(x_{j_{n-k}}) dx_{j_1} \ldots dx_{j_{n-k}}$$

is MRR_2 in the variables $x_{\nu_1}, \ldots, x_{\nu_k}$, where $\{\nu_1, \ldots, \nu_k\}$ and $\{j_1, \ldots, j_{n-k}\}$ are complementary sets of indices $\{1, \ldots, n\}$ with $n \geqq 3$.

There are several recent papers in which reverse regular functions are discussed. See Ebrahimi and Ghosh (1981), and Block, Savits and Shaked (1982).

Let $X_1, \ldots, X_n$ be random variables with joint distribution function F. Let $\Psi_{k_1, \ldots, k_n}(x_1, \ldots, x_n)$ be defined as in section 2. We can introduce the concept of dependence by reverse regular rule as follows. For $k_1, \ldots, k_n \geq 0$, the positive random vector $(X_1, \ldots, X_n)$ is said to be dependent by reverse regular rule of order 2 with degree $(k_1, \ldots, k_n)$, denoted by

$DRR(k_1,\ldots,k_n)$, if $\Psi_{k_1,\ldots,k_n}(x_1,\ldots,x_n)$ is an S-MRR_2 function.

Since $\Psi_{0,\ldots,0}(x_1,\ldots,x_n) = f(x_1,\ldots,x_n)$ if the joint density exists, the condition that a random vector $(X_1,\ldots,X_n)$ is DRR(0,...,0) is equivalent to the condition that the joint density is S-MRR_2 (or RR_2 in n = 2 case). For example, the multinomial distribution, multivariate hypergeometric distribution, etc., are DRR(0,...,0), (see Block et al (1982) and Karlin-Rinott (1980)). Also a random vector $\underset{\sim}{X}$ is DRR(1,...,1) means that $\overline{F}(x_1,\ldots,x_n)$ is an S-MRR_2 function when $n \geq 3$, or is RR_2 when n = 2.

One of the main results in Lee (1983b) is that if a distribution is dependent by reverse regular rule of a certain order, then it is dependent by reverse regular rule for all higher orders. Therefore if a distribution has an S-MRR_2 density, then its joint survival function is again S-MRR_2.

The DRR families are also closed under linear transformations. The joint distribution of two independent sets of DRR random variables is DRR. Assume $(X_1,\ldots,X_n)$ is $DRR(k_1,\ldots,k_n)$ with $k_i = 0$ or 1 for some $1 \leq i \leq n$, then the subvector $(X_1,\ldots,X_{i-1},X_{i+1},\ldots,X_n)$ is $DRR(k_1,\ldots,k_{i-1},k_{i+1},\ldots,k_n)$.

Let $\underset{\sim}{X}$ be a positive random vector. If $\underset{\sim}{X}$ is DRR(1,...,1), then $\underset{\sim}{X}$ is RCSD (see Brindley and Thompson (1972)). If $\underset{\sim}{X}$ is DRR(0,...,0,1), then $\underset{\sim}{X}$ is CDS (see Ebrahimi and Ghosh (1981)).

Random vectors with DRR properties can also be characterized by certain reliability functions as was discussed in section 3. Let $\underset{\sim}{X}$ be a positive absolutely continuous random vector. Then

(1) If $\underset{\sim}{X}$ is DRR(0,...,0,1), then $r(x_n | \underset{\sim}{X}^{(n)} = \underset{\sim}{x}^{(n)})$ is increasing in $\underset{\sim}{x}^{(n)} \in S_{\underset{\sim}{X}}(n)$, for any x_n.

(2) If $\underset{\sim}{X}$ is DRR(0,...,0,2), then $m(x_n|\underset{\sim}{X}^{(n)} = \underset{\sim}{x}^{(n)})$ is decreasing in $\underset{\sim}{x}^{(n)} \in S_{\underset{\sim}{X}(n)}$ for any x_n.

(3) If $\underset{\sim}{X}$ is DRR(0,...,m) for some $m \geq 2$, then

$$\frac{E[(X_n-x_n)^{m-1}|X_n > x_n, \underset{\sim}{X}^{(n)} = \underset{\sim}{x}^{(n)}]}{E[(X_n-x_n)^{m-2}|X_n > x_n, \underset{\sim}{X}^{(n)} = \underset{\sim}{x}^{(n)}]}$$

is decreasing in $\underset{\sim}{x}^{(n)} \in S_{\underset{\sim}{X}^{(n)}}$ for any x_n.

(4) If $\underset{\sim}{X}$ is DRR(1,...,1), then $r(x_j|\underset{\sim}{X}^{(j)} > \underset{\sim}{x}^{(j)})$ is increasing in $\underset{\sim}{x}^{(j)}$, for any x_j, $j = 1,\ldots,n$.

Another conclusion which follows from the above is that if $(X_1,\ldots,X_n)$ is DTP$(k_1,\ldots,k_n)$, and is also DRR$(k_1,\ldots,k_n)$ with the same degrees, then $X_1,\ldots,X_n$ are independent.

REFERENCES

(1) Ahmed, A.N., Langberg, N.A., Leon, R. and Proschan, F. (1987). Two concepts of positive dependence with applications in multivariate analysis. Technical Report M486, Department of Statistics, Florida State University.

(2) Barlow, R.E. and Proschan, F. (1975). *Statistical Theory of Reliability and Life Testing*. Holt, Rinehart and Winston, Inc., New York

(3) Block, H.W. (1977). Monotone Failure rates for multivariate distributions. *Naval Research Logistics Quarterly* 24, 627-637.

(4) Block, H.W., Savits, T.H. and Shaked, M. (1982). Some concepts of negative dependence. *Ann. Probab.* 10. 765-772.

(5) Block, H.W. and Ting, M. (1981). Some concepts of multivariate dependence. *Comm. Statist.* A10(8), 749-762.

(6) Brindley,E.C., Jr. and Thompson, W.A. (1972). Dependence and aging aspects of multivariate survival. *J. Amer. Statist. Assoc.* 67, 822-830.

(7) Ebrahimi, N. and Ghosh, M. (1981). Multivariate negative dependence. *Comm. Statist.*, A10, 307-337.

(8) Esary, J.D. and Marshall, A.W. (1979). Multivariate distributions with increasing hazard rate average. *Ann. Probab.* 7, 359-370.

(9) Esary, J.D. and Proschan, F. (1968). Generating associated random variables. Boeing Scientific Research Laboratories Document D1-82-0696. Seattle, Washington.

(10) Harris, R. (1970). A multivariate definition for increasing hazard rate distributions. *Ann. Math. Statist.* 41, p. 1456-1465.

(11) Johnson, N. and Kotz, S. (1975). A vector multivariate hazard rate. *Journal of Multivariate Analysis*, 5, 53-66.

(12) Karlin, S. (1968). *Total Positivity*, Vol. I. Stanford University Press. Stanford, California.

(13) Karlin, S. and Rinott, Y. (1980). Classes of orderings of measures and related correlation inequalities. II. Multivariate reverse regular rule distributions. *J. Mult. Anal.* 10, 499-516.

(14) Lee, M.-L.T.(1983a) Dependence by total positivity. *Annals of Probability*, to appear.

(15) Lee, M.-L.T.(1983b) Dependence by reverse regular rule. *Annals of Probability*, to appear.

(16) Lehmann, E.L. (1966). Some concepts of dependence. Ann. Math. Statist. 37, 1137-1153.

(17) Shaked, M. (1977). A family of concepts of positive dependence for bivariate distributions. J. Amer. Statist. Assoc. 72, 642-650.

(18) Yanagimoto, T. (1972). Families of positively dependent random variables. Annals of the Institute of Statistical Mathematics 24, 559-573.

Special Session:

U.S. Air Force Perspectives on Failure Models and Reliability–Availability–Maintainability

U.S. AIR FORCE PERSPECTIVES ON FAILURE MODELS AND RELIABILITY/AVAILABILITY/MAINTAINABILITY

INTRODUCTION

Robert T. Smythe, Program Advisor,
Air Force Office of Scientific Research

The process by which basic research becomes operationally useful is often quite long and full of unexpected twists and turns. Although the aim of basic research is not the solution of a particular operational problem, a general familiarity with potential areas of application can be extremely valuable in guiding the researcher's efforts toward ultimately useful ends.

In my capacity as advisor to the Probability and Statistics program of AFOSR, I have repeatedly been asked by researchers in the academic community what the "real problems" of the Air Force are, and what particular research directions might be responsive to these needs. On the other hand, I have found a desire and a need on the part of the Air Force scientists in the laboratories and field commands to better acquaint themselves with recent basic research progress in modeling and inference for reliability/availability/maintainability.

The informal session on US Air Force perspectives, held during the conference on Stochastic Failure Models, Replacement and Maintenance Policies, and Accelerated Life Testing, was an

ISBN 0-12-041420-1

attempt to respond to both of these concerns. Several Air Force scientists were invited to share with the conference attendees (most of them university-based researchers) their ideas on how stochastic modeling might be useful in the problems they face.

Presentations were made by Francis J. O'Meara, Chief, Operations Analysis Division, Office of Science and Research, Strategic Air Command; Anthony J. Feduccia, Chief, Systems Reliability and Engineering Branch, Reliability and Compatibility Division, Rome Air Development Center; and Gary R. Adams, Chief, New Engines Division, Directorate of Engineering, Deputy for Propulsion, Aeronautical Systems Division.

Each speaker was asked to describe briefly his field of responsibility and to indicate areas in which basic research could be valuable. Summaries of the three presentations follow.

PRESENTATION I: Comments to Stochastic Failure Models Conference, *Francis J. O'Meara*, Strategic Air Command.

We are pleased to represent the Strategic Air Command (SAC) at this conference. SAC is the United States Air Force long-range strike force composed of manned bombers, aerial tankers, intercontinental ballistic missiles (ICBMs), air-launched cruise missiles (ALCMs), air-to-surface missiles (ASMs), and a variety of nuclear and conventional munitions.

We are associated with the Office of Science and Research at SAC headquarters located at Offutt AFB, near Omaha, Nebraska. We serve as scientific advisers and technical consultants to the Commander in Chief and have a staff of 19 scientists and engineers to perform our four functions: research, studies, analysis and evaluations. Some of the areas in which our services have been utilized include: requirements for new and improved weapon systems, cost effectiveness studies, force sizing and mix, operational testing and evaluations, training and exercises, estimation of reliability and accuracy, logistics (including scheduling, provisioning and maintenance), plans and tactics, war gaming, and weapon effects.

To get even with you for all the new acronyms and abbreviations you have thrown at us in this conference, we are going to counter with an equation that summarizes much of the mission and concerns of SAC:

$$DE = PLS \times WSR \times PTP \times PD$$

where

DE = Damage Expectancy
PLS = Pre-Launch Survivability (function of enemy initiatives)
WSR = Weapon Systems Reliability (function of system performance)
PTP = Probability to Penetrate (function of enemy defense)
PD = Probability of Damage (function of weapon accuracy, yield, and target damage criteria).

Before presenting some examples of our statistical problems and concerns, allow us to comment on several papers that have been presented at this conference thus far. First, Barlow's paper "A Survey of Network Reliability" touched on one of the most important and pressing issues of our time, connectivity. Since network complexity does, indeed, grow in an exponential manner, any effort such as Dr. Barlow's to simplify network representation has a tremendous payoff to SAC and the Air Force. A number of papers have been presented on damage from shock waves and the recovery to a normal state, mostly in the context of a stochastic process relating to reservoirs. We see this work in a different light. Consider an enemy trying to pin down our missile force with a nuclear attack. We are then interested in looking for periods of time when the transient weapon effects disappear, creating fire-out windows within which to retaliate with our own missiles.

The following three examples are illustrative of our major statistical problems and issues at this time:

Estimation of the Impact of Reliability on Mission Effectiveness

Given a complex system of many components and varying failure rates, estimate the best strategy for the investment of funds to improve damage expectancy (DE) through increased reliability. Now what complicates this problem is that many of our systems do not simply work or not, but fail in such a way as to cause degraded operations. For example, radar altimeter failure may cause us to climb to a higher flight level which in turn lowers our probability to penetrate (PTP) due to enemy action. Dr. Ismail Shimi, of OSR, was particularly interested in these partial failure modes before his untimely death.

Maintenance During Extended Operations

Given a fixed number of aircraft, crew members, maintenance personnel, demand for sorties, limitation on sortie rates and a maintenance policy, how many spares should we take with us to guarantee a certain level of service for a fixed period of time? This is further complicated by the introduction of cannibalization and attrition due to enemy action.

Estimation of ICBM Weapon System Reliability (WSR) and Accuracy

How many "extra" ICBMs should SAC buy for operational testing to obtain an initial estimate of WSR/accuracy and then to monitor these factors over the life of the weapon system? Some factors to be considered include the cost of test assets, the cost of testing, the size of the deployed force, and the relative importance of the system's contribution to our strategic war effort. There is extreme pressure to reduce the escalating costs of acquiring and operating new weapon systems. Using "conventional statistical procedures" the number of ICBMs expended in destructive testing to obtain estimates of reliability and accuracy could easily exceed the deployed force in numbers and cost. We need methods of obtaining good estimates more cheaply. Some possibilities include making more use of ground tests and simulations, making better use of similar information from other systems, using engineering estimates and relying more on component level as opposed to system level testing. We would like to leave you with a challenge -- "How can we combine information from a number of sources, collected over a period of time, into a single good estimate?"

PRESENTATION II: Reliability and Maintainability Research - An RADC Point of View, *Anthony J. Feduccia,* RADC/RBE.

Traditionally, reliability and maintainability (R&M) research has centered on the development of new, sometimes unique, mathematical and statistical expressions aimed at solving a particular problem under definite assumptions. We have witnessed, for example, the development of system reliability models based on nonexponential part failure distributions; reliability demonstration techniques for these models; and, a proliferation of different Markov models for evaluation of redundant network reliability and availability. Indeed, some exciting research findings have occurred which have been major contributors to increased system reliability and maintainability. But, the emphasis in military R&M is changing. More and more attention is being devoted to increasing system operational readiness and logistics supportability - and rightly so, since the services are all suffering from decreased maintenance manpower and skill levels while, at the same time, system complexity and performance requirements are increasing. The Air Force research community must respond to this need by focussing attention on studies directly related to Air Force system needs.

The traditional reliability prediction technique, which assumed only part failures caused system failure, is outmoded and is being replaced by "total system" availability methodologies. Schemes need to be developed to account for system failure as a result of software errors, human maintenance and operator errors. The effect of preventive and corrective maintenance and the impact of the maintenance concept and logistics environment under operational conditions must be considered early in the design of the system. We need models and guidelines to use for this purpose.

Reliability testing is changing too. Only rarely do we see reliability qualification testing performed using the sequential tests or fixed length tests found in MIL-STD-781. Instead, test-analyze-and-fix (TAAF) testing and reliability growth tests are being used. We need high confidence demonstration methods and procedures tied to or associated with reliability growth tests for use in conjunction with (or as a replacement for) short-term reliability qualification tests. Also, demonstration techniques are needed to demonstrate system MTBF based on system components which are tested separately for different numbers of hours. Additionally, the development of physically valid accelerated tests should be pursued to develop failure rates for component parts, especially the new technology VLSIC devices.

Equipment/system testability is emerging as an engineering entity of its own. Testability is defined as a systems' capability to detect and isolate failures within its functions. RADC's testability program has made great strides towards the ultimate objectives of having the capability to specify, predict and demonstrate testability, but much more work needs to be done in designing "SMART" build-in-test (BIT) to reduce the number of unnecessary removals; in applying artificial intelligence techniques to improving maintenance; and, in developing optimization procedures for test system development. This includes cost/performance trade-off analyses considering the number of maintenance levels needed, skill levels required, sparing etc. for maximum payoff in effectiveness and cost.

Other areas of needed research exist but the bottom line is that 1984 research must be "system relevant" and particularly aimed at increasing system readiness and reducing costs.

PRESENTATION III: Application of Stochastic Failure Models to Gas Turbine Engines, *Gary R. Adams*, Aeronautical Systems Division.

Aeronautical Systems Division (ASD) is part of the Air Force Systems Command (AFSC) and is responsible for the development, acquisition and deployment of manned weapons systems for the Air Force. The Deputy for Propulsion is in turn responsible for the development and acquisition of gas turbine engines. Field support of the engines is provided by the Air Force Logistics Command (AFLC) supplemented by engineering and management assistance provided by ASD. Both ASD and AFLC Headquarters are located at Wright-Patterson Air Force Base in Dayton, Ohio, providing for close coordination between the two commands. The importance of this coordination can be seen in Figures 1 and 2.

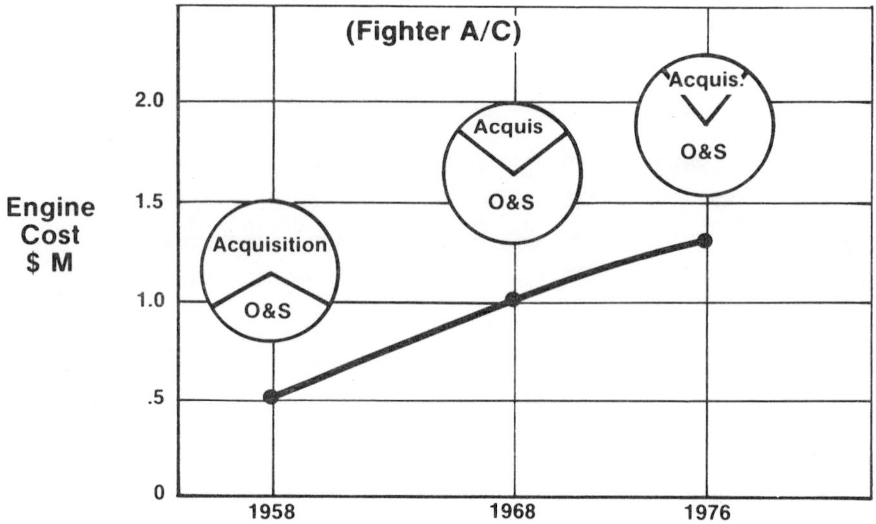

FIGURE 1. Acquisition Cost vs Operation & Support Cost.

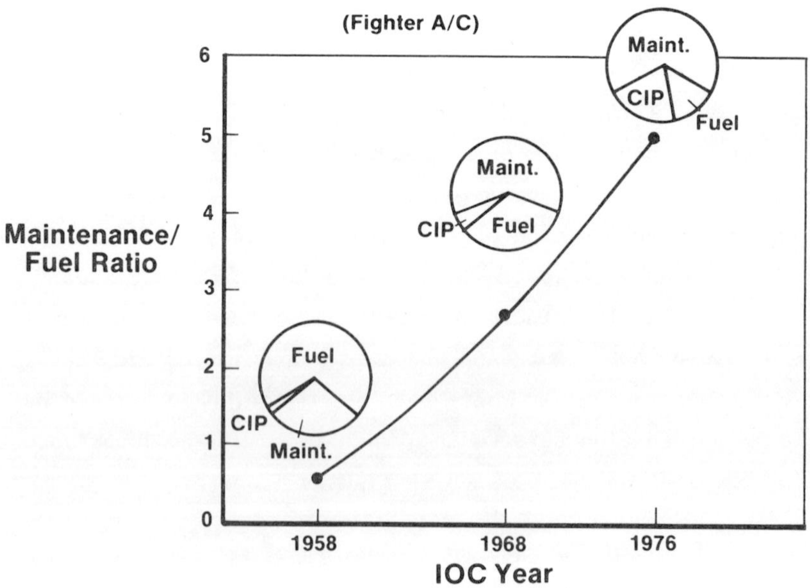

FIGURE 2. Maintenance Cost Drives O&S.

In the 1960 time period, the total cost of ownership of fighter systems was dominated by acquisition cost. Today, the operating and support cost (O&S) is the bulk of the total cost. Obviously, the increase of fuel costs forms a part of this change, but if we examine the ratio of maintenance costs to fuel costs, we see that maintenance costs have actually grown at a faster rate. This is primarily due to the increased complexity and sophistication of these systems. Additionally, the amount of money being spent on the Component Improvement Program (CIP) is increasing. CIP represents continued engineering development to improve the capabilities of the weapons systems and to address service revealed problems.

Gas turbine engines provide a number of challenging problems. The engine itself is an extremely complex mechanical system even when we ignore the aerodynamic aspects of the machinery. In large engines, the compressors and turbines rotate at speeds of 10,000 to 14,000 RPM while small engines run at speeds in the 60,000 RPM range. And the rotating components typically operate with clearance in the range of 10 to 40 thousandths of an inch. The combustor provides a very high heat release capability in a small volume and the overall power density of the gas turbine engine is one of its major assets. The mechanical complexity is further aggravated by the high temperatures and thermal cyclic characteristics of gas turbines. Gas path temperatures of 2000°F are common, and typical transients from idle to full power are accomplished in 5-10 seconds. The combination of thermal stresses, large changes in operating pressures and centrifugal forces results in a very complex set of stresses acting on the various components.

Mounted on the outside of the engine are a number of control and accessory components. The parts count of the components is about the same as the parts count of the engine itself, and not surprisingly, half the maintenance is driven by these components.

These components include hydraulic, pneumatic, electrical, and mechanical systems. Our newer engines have engine mounted digital control and monitoring systems which are exposed to extremely difficult environmental conditions and take special packaging, including fuel cooling and vibration isolation, in order to survive. Overall, gas turbine engines encompass a broad spectrum of engineering fields and represent a very difficult failure analysis/projection problem. Typical failure modes experienced in gas turbine engines which are good candidates for stochastic modelling are:

- (i) Bearings/Gears/Shafts
- (ii) Engine hot section
  - (a) Combustor
  - (b) Turbine blades and vanes
- (iii) Disks
- (iv) Controls and Accessories
  - (a) Mechanical
  - (b) Electrical

Actual applications can be considered in three areas; production, operations, and maintenance.

Some of the production problems which can be addressed are:

- (i) Component testing requirements
- (ii) Sampling plan
  - (a) Production quality/infant mortality
  - (b) Fleet wide MTBF
  - (c) Warranty/guarantee

Part of our inherent problem is that a major program will only generate a production rate of 30 to 50 articles per month. And gas turbine engines are very labor intensive, so that extensive testing and inspection at the final stages is very costly. On the other hand, once delivered to the field, infant mortality is also very costly due to the high transportation costs, spares requirements, and labor intensive maintenance. The ability to analyze and project failure rates would be very beneficial in determining the optimum level of testing at the piece part, component and full engine levels. Warranties are one way of motivating the contractor to improve quality, but it also requires similar failure modeling capabilities to define a good warranty. A good warranty is one where both parties share the risk of achieving the desired capabilities.

For the operational phase, failure modeling can be applied to two general areas, flight safety and mission readiness. For flight safety, hazardous events analysis/projection is crucial. Obviously more than pure cost is involved. Current methods are based on Failure Modes and Effects analysis in conjunction with fault trees to determine the hazards of particular failure modes. Rate estimates are required to provide the final risk assessment. The ability of the weapons system to meet its mission is of upmost importance, and generally two figures of merit are used. Availability is the probability that the system is mission capable when desired. Mission Reliability is the probability that once commenced, the system can complete its operational mission. The application of stochastic models to these problems should be obvious.

Before addressing the maintenance aspects, we'll examine the current Air Force Maintenance policy. Reliability Centered Maintenance (RCM) is the Air Force application of On Condition Maintenance (OCM). Hard time limits are used for wear out failure modes based on economic considerations of secondary damage versus replacement cost. It is also used to control risk in safety critical parts. Since an engine may have a large number of time limited parts (typically 100 different part numbers) each of which have different life limits, the maintenance policy required for parts replacement is very complex. Parts must be grouped into convenient groups which can be replaced at a single time limit in order to minimize the total time an engine is out of service. The generation of an optimum scheduled maintenance plan requires the application of a number of stochastic models. On Condition Maintenance is dependent upon two general criteria. The failure modes must not be safety critical and the failure mode must have a conditional test which can be applied to determine what maintenance is required. These conditional tests can range from built-in-test for electronics to visual inspections. Conditional Maintenance is generally applied to non-critical consumables. Whenever the engine is down for repairs, these components will be inspected and replaced as required. This is referred to as opportunistic maintenance.

Application of stochastic modeling to the maintenance/logistics problem would be confined to the first two categories of maintenance, hard time limits and on condition maintenance, respectively. For Hard Time Limits, modeling could be used to define the initial maximum operating time limits (MOH) and subsequently used to extend these limits as a better data base becomes available. It is important to note that the term "Time" is used in a generic sense, and does not necessarily mean operating hours. Rather, time is a measure of damaging events which could be thermal cycles. In actuality each part experiences different levels of damage with respect to both the steady state and cyclic operation of the engine. In general, a usage profile is estimated and "time" limits defined as a function of operating hours, cyclic operation or some combination of these factors.

For On Condition Maintenance, stochastic modeling could be used to determine optimum spares provisioning for unscheduled maintenance. It would have to consider operational availability and time to repair in addition to failure rates. Since many repairs are accomplished at a remote depot, it also must consider the pipeline requirements. Stochastic modeling may also have an application in better defining conditional limits. In most cases, the actual condition is inferred by indirect measurements resulting in some ambiguity. Conditional limits must consider false alarm rates (generation of unnecessary maintenance) as well as the ability to recognize actual maintenance requirements.

In summary, there are a number of areas where stochastic models could be applied (or improved) for gas turbine engines. And, while most Air Force project engineers understand the basics of stochastic principles, they generally do not have the expertise or the time to develop the basic tools required for our applications. A significant opportunity exists for generating the procedures required to apply stochastic modeling to the development, acquisition and support of gas turbine engines within the Air Force. And while many of these procedures may have broader application, it is important to make them specific enough to be readily recognizable by the project engineers who can use them, and easy enough to be applied by engineers who do not have strong statistical backgrounds.